NOTES ON FORCING AXIOMS

LECTURE NOTES SERIES
Institute for Mathematical Sciences, National University of Singapore

Series Editors: Chitat Chong and Wing Keung To
Institute for Mathematical Sciences
National University of Singapore

ISSN: 1793-0758

Published

*For the complete list of titles in this series, please go to
http://www.worldscientific.com/series/LNIMSNUS

Lecture Notes Series, Institute for Mathematical Sciences,
National University of Singapore
Vol.
26

NOTES ON FORCING AXIOMS

Stevo Todorcevic
University of Toronto, Canada

Editors

Chitat Chong
Qi Feng
Yue Yang
National University of Singapore, Singapore

Theodore A Slaman
W Hugh Woodin
University of California, Berkeley, USA

NEW JERSEY · LONDON · SINGAPORE · BEIJING · SHANGHAI · HONG KONG · TAIPEI · CHENNAI

Published by

World Scientific Publishing Co. Pte. Ltd.

5 Toh Tuck Link, Singapore 596224

USA office: 27 Warren Street, Suite 401-402, Hackensack, NJ 07601

UK office: 57 Shelton Street, Covent Garden, London WC2H 9HE

Library of Congress Cataloging-in-Publication Data
Todorcevic, Stevo.
    Notes on forcing axioms / by Stevo Todorcevic (University of Toronto, Canada) ; edited by
Chitat Chong (National University of Singapore, Singapore), Qi Feng (National University of
Singapore, Singapore), Yue Yang (National University of Singapore, Singapore), Theodore A.
Slaman (University of California, Berkeley, USA), & W. Hugh Woodin (University of California,
Berkeley, USA).
        pages cm. -- (Lecture notes series (Institute for Mathematical Sciences, National University of
Singapore) ; volume 26)
    Includes bibliographical references and index.
    ISBN 978-9814571579 (hardcover : alk. paper)
    1. Forcing (Model theory) 2. Axioms. 3. Baire classes. I. Chong, C.-T. (Chi-Tat), 1949– editor.
II. Feng, Qi, 1955– editor. III. Yang, Yue, 1964– editor. IV. Slaman, T. A. (Theodore Allen),
1954– editor. V. Woodin, W. H. (W. Hugh), editor. VI. Title.
    QA9.7.T63 2014
    511.3--dc23

                                                                            2013042520

British Library Cataloguing-in-Publication Data
A catalogue record for this book is available from the British Library.

Printed in Singapore

Contents

Foreword
by Series Editors

The Institute for Mathematical Sciences (IMS) at the National University of Singapore was established on 1 July 2000. Its mission is to foster mathematical research, both fundamental and multidisciplinary, particularly research that links mathematics to other efforts of human endeavor, and to nurture the growth of mathematical talent and expertise in research scientists, as well as to serve as a platform for research interaction between scientists in Singapore and the international scientific community.

The Institute organizes thematic programs of longer duration and mathematical activities including workshops and public lectures. The program or workshop themes are selected from among areas at the forefront of current research in the mathematical sciences and their applications.

Each volume of the *IMS Lecture Notes Series* is a compendium of papers based on lectures or tutorials delivered at a program/workshop. It brings to the international research community original results or expository articles on a subject of current interest. These volumes also serve as a record of activities that took place at the IMS.

We hope that through the regular publication of these *Lecture Notes* the Institute will achieve, in part, its objective of reaching out to the community of scholars in the promotion of research in the mathematical sciences.

September 2013

Chitat Chong
Wing Keung To
Series Editors

Foreword
by Volume Editors

The series of Asian Initiative for Infinity (AII) Graduate Logic Summer School was held annually from 2010 to 2012. The lecturers were Moti Gitik, Denis Hirschfeldt and Menachem Magidor in 2010, Richard Shore, Theodore A. Slaman, John Steel, and W. Hugh Woodin in 2011, and Ilijas Farah, Ronald Jensen, Gerald E. Sacks and Stevo Todorcevic in 2012. In all, more than 150 graduate students from Asia, Europe and North America attended the summer schools. In addition, two postdoctoral fellows were appointed during each of the three summer schools. These volumes of lecture notes serve as a record of the AII activities that took place during this period.

The AII summer schools was funded by a grant from the John Templeton Foundation and partially supported by the National University of Singapore. Their generosity is gratefully acknowledged.

October 2013

Chitat Chong
Qi Feng*
Yue Yang
National University of Singapore, Singapore

Theodore A. Slaman
W. Hugh Woodin
University of California at Berkeley, USA

Volume Editors

---

*Current address: Chinese Academy of Sciences, China.

Preface

Baire category method as a tool for showing the existence of interesting mathematical objects is well established in mathematics. The set-theoretic technique of Forcing brings this method to another level of sophistication and potential applicability. The purpose of the notes is to expose some of these.

This set of notes was build over the last ten years and tested on courses I gave in Paris (Spring of 2003), Toronto (Spring of 2005 and Fall of 2012), and Singapore (Summer of 2012).[1]

I would like to thank the students who took the courses for the help in organizing this set of lecture notes.

Stevo Todorcevic
Spring, 2013

---

[1]The Singapore course was a part of the AII Graduate Summer School jointly organized and funded by the John Templeton Foundation and the Institute for Mathematical Sciences of the National University of Singapore. I would like to thank these institutions for their support.

Chapter 1

Baire Category Theorem and the Baire Category Numbers

1.1 The Baire category method – a classical example

Recall the following:

Theorem 1 (Baire Category Theorem). *Given any compact Hausdorff space, or complete metric space X, the intersection of any countable family of dense open subsets of X is dense in X. In particular, such an intersection is always nonempty.*

Theorem 2 (K. Weierstrass). *There is a continuous nowhere differentiable function on the closed unit interval $[0,1]$.*

Proof. (S. Banach). Let $\mathcal{C}$ denote the set of all continuous functions on $[0,1]$. Note that $\mathcal{C}$ is a Banach space under the uniform norm

$$\|f\|_\infty = \sup\{|f(x)| : x \in [0,1]\}.$$

It can easily be shown that $\mathcal{C}$ is a separable complete metric space under the metric

$$\rho(f,g) = \|f - g\|_\infty = \sup\{|f(x) - g(x)| : x \in [0,1]\}.$$

For each $n > 1$ define

$$F_n = \left\{ f \in \mathcal{C} : (\exists x \in \left[0, 1 - \frac{1}{n}\right])(\forall 0 < h < 1 - x)\,|f(x+h) - f(x)| \le nh \right\}.$$

Claim. *Each F_n is closed in $\mathcal{C}$.*

Proof of claim. Suppose that $f \in \overline{F_n}$. Then there is a sequence $\{f_k\}_{k<\omega}$ of functions in F_n converging uniformly to f. For each $k < \omega$ there is an $x_k \in [0, 1 - \frac{1}{n}]$ witnessing that $f_k \in F_n$. As $\{x_k\}_{k<\omega}$ is a sequence in the compact set $[0, 1 - \frac{1}{n}]$, it has a convergent subsequence, so without loss of generality we may assume that $\{x_k\}_{k<\omega}$ converges to some $x \in [0, 1 - \frac{1}{n}]$.

Suppose that $0 < h < 1 - x$, and $\epsilon > 0$. We wish to show that $|f(x + h) - f(x)| \leq nh + \epsilon$, so that we may conclude that $|f(x+h) - f(x)| \leq nh$, and $f \in F_n$. As f is continuous, there is a $\delta > 0$ such that

$$|x - y| < \delta \;\rightarrow\; |f(x) - f(y)| < \epsilon; \tag{1.1}$$
$$|x + h - y| < \delta \;\rightarrow\; |f(x + h) - f(y)| < \epsilon. \tag{1.2}$$

Then there is a $K \in \mathbb{N}$ such that for any $k \geq K$ we have the following:

(1) $0 < h < 1 - x_k$, and

(2) $\rho(f, f_k) < \frac{\epsilon}{4}$, and

(3) $|x - x_k| < \delta$.

We then have

$$
\begin{aligned}
|f(x+h) - f(x)| &\leq |f(x+h) - f(x_k + h)| + |f(x_k + h) - f_k(x_k + h)| \\
&\quad + |f_k(x_k + h) - f_k(x_k)| + |f_k(x_k) - f_k(x)| \\
&\quad + |f_k(x) - f(x)| \\
&\leq \frac{\epsilon}{4} + \rho(f, f_k) + nh + \rho(f_k, f) + \frac{\epsilon}{4} \\
&\leq nh + \epsilon. \qquad\qquad \square
\end{aligned}
$$

Claim. *Each F_n is nowhere-dense in $\mathcal{C}$.*

Sketch of proof of claim. Suppose that $U \subseteq \mathcal{C}$ is open, and $f \in U$. We wish to find an element $g \in U$ which is not in F_n. This is technical, but can basically be broken down into a few steps.

Step 1: As the piecewise-linear functions are dense in $\mathcal{C}$, we may assume without loss of generality that f is piecewise linear. Say

$$
f(x) = \begin{cases}
m_1 x + b_1, & \text{if } 0 = x_0 \leq x \leq x_1 \\
m_2 x + b_2, & \text{if } x_1 \leq x \leq x_2 \\
\;\vdots \\
m_k x + b_k, & \text{if } x_{k-1} \leq x \leq x_k = 1.
\end{cases}
$$

Step 2: We may also assume without loss of generality that $U = \{g \in \mathcal{C} : \rho(f, g) \leq \epsilon\}$ for some $\epsilon > 0$.

Step 3: For each $i = 1, \ldots, k$ let $\varphi_i(x)$ be a "sawtooth" function on the closed interval $[x_{i-1}, x_i]$ with $\varphi_i(x_{i-1}) = 0 = \varphi_i(x_i)$. Also ensure that $|\varphi_i(x)|$ has maximum value $< \epsilon$, and also each linear piece of φ_i has slope $\pm r_i$ where $|m_i + r_i|, |m_i - r_i| > n$. For x outside of the closed interval $[x_{i-1}, x_i]$ define $\varphi_i(x) = 0$. Let $\varphi(x) = \sum_{i=1}^{k} \varphi_i(x)$.

Step 4: Define $g(x) = f(x) + \varphi(x)$. It is easy to check that $\rho(g, f) < \epsilon$, as $|\varphi(x)| < \epsilon$. Finally, by choice of the r_i, it can be shown that for each $x \in [0, 1 - \frac{1}{n}]$ there is an $0 < h < 1 - x$ such that $|g(x+h) - g(x)| > nh$. $\square$

By the Baire Category Theorem it follows that $\bigcup_{n>1} F_n \neq \mathcal{C}$, and thus there is an $f \in \mathcal{C} \setminus \bigcup_{n>1} F_n$. For each $x \in [0, 1)$, any $M > 0$, and any $\epsilon > 0$ it can be shown that there is a $0 < h < \epsilon$ such that $\frac{|f(x+h) - f(x)|}{h} > M$. This immediately implies that $\lim_{h \to 0^+} \frac{f(x+h) - f(x)}{h}$ does not exist, thus, f is nowhere differentiable. $\square$

Fact. *Slight alterations in the above will show that there is a nowhere differentiable periodic function on $\mathbb{R}$ with period 1.*

1.2 Baire category numbers

Definition. Suppose that X is a topological space (usually a compact Hausdorff space, or a Čech-complete space). We define the *category number* of X to be
$$\mathfrak{m}(X) = \min\{|\mathcal{U}| : \mathcal{U} \subseteq \mathcal{DO}(X), \bigcap \mathcal{U} = \emptyset\}.$$

Definition. Suppose that $\mathbf{K}$ is a class of topological spaces (usually a class of compact Hausdorff spaces or Čech-complete spaces). We define the *category number* of $\mathbf{K}$ to be
$$\mathfrak{m}(\mathbf{K}) = \min\{\mathfrak{m}(X) : X \in \mathbf{K}\}.$$

Notation. Denote the class of all compact Hausdorff spaces satisfying the c.c.c. by $\mathbf{CCC}$. We denote the category number $\mathfrak{m}(\mathbf{CCC})$ by $\mathfrak{m}$. It is easy to show that $\aleph_1 \leq \mathfrak{m} \leq \mathfrak{c}$. The "axiom" (or assumption) "$\mathfrak{m} = \mathfrak{c}$" is called *Martin's Axiom* (MA).

Martin's Axiom has many interesting consequences. For many examples of these, see [8].

There are many classes $\mathbf{K}$ of topological spaces whose category numbers are worth analysing. Among these (in order of inclusion) are

(1) The class of separable completely metrizable spaces. For this class, $\mathfrak{m}(\mathbf{K})$ is otherwise known as the *covering of category*, defined by

$$\mathfrak{m}(\mathbb{R}) = \min\{|\mathcal{A}| : \mathcal{A} \text{ is a family of meagre subsets of } \mathbb{R} \text{ and } \bigcup \mathcal{A} = \mathbb{R}\}.$$

(2) The class $\mathbf{K}_{\text{sep}}$ of separable Čech-complete spaces. For this class, $\mathfrak{m}(\mathbf{K}_{\text{sep}})$ is otherwise denoted $\mathfrak{p}$, and had been already studied combinatorially by F. Rothberger in the 1940's as the minimal cardinal θ with the property that there is a family $\mathcal{F} \subseteq [\mathbb{N}]^{\infty}$ of infinite subsets of $\mathbb{N}$ such that $|\mathcal{F}| = \theta$ and $|\bigcap \mathcal{F}_0| = \aleph_0$ for all finite $\mathcal{F}_0 \subseteq \mathcal{F}$, but there is no infinite $M \subseteq \mathbb{N}$ such that $M \subseteq^* N$ for all $N \in \mathcal{F}$, where $\subseteq^*$ denotes the relation of inclusion modulo a finite set.

(3) The class $\mathbf{K}_{\text{ccc}}$ of c.c.c. Čech-complete spaces. For this class $\mathfrak{m}(\mathbf{K}_{\text{ccc}}) = \mathfrak{m} = \mathfrak{m}(\mathbf{CCC})$.

(4) Let $\mathbf{M}$ be a maximal class for which this notion is nontrivial, i.e. for every compact or Čech-complete space X with no isolated points not belonging to $\mathbf{M}$, we have that $\mathfrak{m}(X) = \omega_1$. For this class $\mathbf{M}$ we will denote $\mathfrak{m}(\mathbf{M})$ by $\mathfrak{mm}$. The assumption that $\mathfrak{mm} > \aleph_1$ is called *Martin's Maximum (MM)*.

It is easy to see that the mapping $\mathbf{K} \mapsto \mathfrak{m}(\mathbf{K})$ is non-increasing, and so we have
$$\aleph_1 \leq \mathfrak{mm} \leq \mathfrak{m} \leq \mathfrak{p} \leq \mathfrak{m}(\mathbb{R}) \leq \mathfrak{c}.$$

1.3 $\mathcal{P}$-clubs

In this section, we explain the class of posets whose category number will give us another formulation of the invariant $\mathfrak{mm}$ introduced above. Central to this goal is the notion of a $\mathcal{P}$-club.

Definition. Given a partial order $\mathcal{P}$, a *$\mathcal{P}$-club* is a family $\langle C(p) : p \in \mathcal{P} \rangle$ of subsets of ω_1 satisfying the following conditions:

1. *($\mathcal{P}$-monotonicity):* $p \leq q$ implies $C(p) \supseteq C(q)$, and

2. *($\mathcal{P}$-closedness):* for each $\gamma \in \omega_1 \setminus C(p)$ there is a $\beta < \gamma$ and a $q \leq p$ such that $C(r) \cap (\beta, \gamma] = \emptyset$ for all $r \leq q$, and

3. *($\mathcal{P}$-unboundedness):* for each $\alpha < \omega_1$ there is a $q \leq p$ and a $\gamma \in C(q)$ with $\gamma \geq \alpha$.

The following fact is immediate from the above definition:

Fact. *If $\langle C(p) : p \in \mathcal{P} \rangle$ is a $\mathcal{P}$-club, then $C(p)$ is closed for each $p \in \mathcal{P}$.*

Proof. If $\gamma \notin C(p)$, then by $\mathcal{P}$-closedness there is a $\beta < \gamma$ and a $q \leq p$ such that $C(r) \cap (\beta, \gamma] = \emptyset$ for all $r \leq q$. In particular, $C(r) \cap (\beta, \gamma] = \emptyset$, and thus by $\mathcal{P}$-monotonicity it follows that $C(p) \cap (\beta, \gamma] = \emptyset$, and so γ is not the limit of any increasing sequence in $C(p)$. $\qquad\square$

Inserting some forcing into this discussion, a $\mathcal{P}$-club is essentially an encoding of a $\mathcal{P}$-name for a club in ω_1. That is, given a $\mathcal{P}$-club $\langle C(p) : p \in \mathcal{P}\rangle$, the family $\dot{C} = \{\langle \check{\alpha}, p\rangle : p \in \mathcal{P}, \alpha \in C(p)\}$ can easily be seen to be a $\mathcal{P}$-name for a closed and unbounded subset of ω_1.

We will now point out a particular family of partial orders which will be extremely important in the sequel:

Definition. A partial order $\mathcal{P}$ is said to *preserve stationary subsets of ω_1*, (or, in short, to be *stationary set preserving*) if for each $\mathcal{P}$-club $\langle C(p) : p \in \mathcal{P}\rangle$ and every stationary set $E \subseteq \omega_1$, the set

$$\{p \in \mathcal{P} : C(p) \cap E \neq \emptyset\}$$

is dense in $\mathcal{P}$.

Lemma. *If a partial order $\mathcal{P}$ does not preserve stationary sets, then there is a $p_0 \in \mathcal{P}$, and a family $\{D_\xi\}_{\xi < \omega_1}$ of dense sets in $\mathcal{P}$ such that for each filter $\mathcal{G}$ in $\mathcal{P}$ with $p_0 \in \mathcal{G}$ there is a $\xi < \omega_1$ such that $D_\xi \cap \mathcal{G} = \emptyset$.*

Proof. Let $\langle C(p) : p \in \mathcal{P}\rangle$ be a $\mathcal{P}$-club, and let $S \subseteq \omega_1$ be a stationary set such that

$$\{p \in \mathcal{P} : C(p) \cap S \neq \emptyset\}$$

is not dense in $\mathcal{P}$. Let $p_0 \in \mathcal{P}$ be such that $C(q) \cap S = \emptyset$ for all $q \leq p_0$.

For each $\gamma < \omega_1$ we define the following two subsets of $\mathcal{P}$:

$$D_\gamma = \{p \in \mathcal{P} : p \perp p_0, \text{ or } p \leq p_0 \text{ and } C(p) \nsubseteq \gamma\},$$

$$E_\gamma = \left\{ p \in \mathcal{P} : \begin{array}{l} p \perp p_0, \text{ or } p \leq p_0 \text{ and either } \gamma \in C(p) \text{ or} \\ (\exists \beta < \gamma)(\forall r \leq p)(C(p) \cap (\beta, \gamma] = \emptyset) \end{array} \right\}.$$

Claim. *Each of the sets D_γ and E_γ is dense-open in $\mathcal{P}$.*

Proof of claim. $\mathcal{P}$-unboundedness immediately implies that D_γ is dense below p_0, and it easily follows that D_γ is dense in $\mathcal{P}$. By $\mathcal{P}$-monotonicity, it also follows that D_γ is open.

Similarly, $\mathcal{P}$-closedness immediately implies that E_γ is dense below p_0, and it easily follows that E_γ is dense in $\mathcal{P}$. $\mathcal{P}$-monotonicity and the definition of E_γ easily give us that E_γ is open. $\square$

Suppose that $\mathcal{G}$ is a filter in $\mathcal{P}$ with $p_0 \in \mathcal{G}$ which intersects each D_γ and each E_γ. Let $C = \bigcup_{p \in \mathcal{G}} C(p)$.

Claim. *C is a club in ω_1.*

Proof of claim. Suppose that $\{\alpha_n\}_{n \in \omega}$ is an increasing sequence in C, and let $\gamma = \sup_{n \in \omega} \alpha_n$. For each $n \in \omega$ choose some $q_n \in \mathcal{G}$ with $\alpha_n \in C(q_n)$. As $\mathcal{G} \cap E_\gamma \neq \emptyset$, let $p \in \mathcal{G} \cap E_\gamma$. Note that $p \leq p_0$, as $p_0 \in \mathcal{G}$. Then either $\gamma \in C(p)$, or $(\exists \beta < \gamma)(\forall r \leq p)(C(r) \cap (\beta, \gamma] = \emptyset)$. If $\gamma \notin C(p)$, pick $\beta < \gamma$

such that $C(r) \cap (\beta, \gamma] = \emptyset$ for all $r \leq p$. Then there is an $n \in \omega$ with $\beta < \alpha_n$. Note that p and q_n are compatible in $\mathcal{G}$, so choose some $r \in \mathcal{G}$ with $r \leq p, q_n$.

As $r \leq p$, by choice of p we have $C(r) \cap (\beta, \gamma] = \emptyset$, however, as $r \leq q_n$, by choice of q_n and $\mathcal{P}$-monotonicity we have

$$C(r) \cap (\beta, \gamma] \supseteq C(r) \cap [\alpha_n, \gamma] \supseteq C(q_n) \cap [\alpha_n, \gamma] \neq \emptyset,$$

which is a contradiction. Therefore $\gamma \in C(p) \subseteq C$, and so C is closed.

For any $\gamma < \omega_1$, as $\mathcal{G} \cap D_\gamma \neq \emptyset$ and $p_0 \in \mathcal{G}$, we immediately have that $C \not\subseteq \gamma$, and therefore C is unbounded in ω_1. □

Note that for each $p \in \mathcal{G}$, as $\mathcal{G}$ is a filter and $p_0 \in \mathcal{G}$, there is a $q \in \mathcal{G}$ with $q \leq p, p_0$. As $q \leq p$, by $\mathcal{P}$-monotonicity $C(q) \supseteq C(p)$. As $q \leq p_0$, by choice of p_0 we have $C(q) \cap S = \emptyset$. Therefore

$$C(p) \cap S \subseteq C(q) \cap S = \emptyset.$$

From this it follows that $C \cap S = \emptyset$, contradicting the fact that S is a stationary set. □

1.4   Baire category numbers of posets

We will now give a formulation of the Baire category numbers $\mathfrak{p}$, $\mathfrak{m}$ and $\mathfrak{mm}$ using partially ordered sets rather than topological spaces.

Definition. Given a partial order $\mathcal{P}$, we define its *category number* as follows:

$$\mathfrak{m}(\mathcal{P}) = \min \left\{ |\mathcal{D}| : \begin{array}{l} \mathcal{D} \text{ is a family of dense-open subsets of } \mathcal{P} \text{ and} \\ \text{there is no filter in } \mathcal{P} \text{ which meets each set in } \mathcal{D} \end{array} \right\}.$$

Fact. $\mathfrak{m}(\mathcal{P}) \geq \omega_1$ *for every partial order* $\mathcal{P}$.

Notation. $\mathfrak{mm} = \min\{\mathfrak{m}(\mathcal{P}) : \mathcal{P} \text{ is stationary set preserving}\}$.

Theorem 3. *Every c.c.c. partial order preserves stationary sets.*

Proof. Let $\mathcal{P}$ be a c.c.c.-poset, $\langle C(p) : p \in \mathcal{P} \rangle$ a $\mathcal{P}$-club, and let $S \subseteq \omega_1$ be a given stationary set. Let $\bar{p} \in \mathcal{P}$ be a given condition. Take a countable elementary submodel $\mathcal{M} \prec \langle H_\theta, \in \rangle$ for some large enough θ containing $\mathcal{P}, \langle C(p) : p \in \mathcal{P} \rangle, S, \bar{p}$ and such that $\delta = \omega_1 \cap \mathcal{M} \in S$.

By $\mathcal{P}$-unboundedness there is a $q \leq \bar{p}$ such that $C(q) \not\subseteq \delta$.

Claim 3.1. $\delta \in C(q)$.

Proof of claim. If not, then by $\mathcal{P}$-closedness there is a $\beta < \delta$ and a $\bar{r} \leq q$ such that $C(s) \cap (\beta, \delta] = \emptyset$ for all $s \leq \bar{r}$. Note that as $\beta \in \mathcal{M}$, then the set $\mathcal{X} = \{r \leq \bar{p} : C(s) \not\subseteq \beta + 1\}$ is in $\mathcal{M}$. Also note that $\bar{r} \in \mathcal{X}$.

Suppose that $\bar{r}$ is not compatible with any member of $\mathcal{X} \cap \mathcal{M}$. Let $\mathcal{X}_0 \in \mathcal{M}$ be such that $\mathcal{M} \models$ "$\mathcal{X}_0$ is a countable subset of $\mathcal{X}$". By elementarity it follows that $\mathcal{X}_0$ is a countable subset of $\mathcal{X}$, and therefore $\mathcal{X}_0 \subseteq \mathcal{M}$. In particular we know that $\bar{r} \notin \mathcal{X}_0$. As $r \perp \bar{r}$ for each $r \in \mathcal{X}_0$ by assumption, it follows that $H_\theta \models (\exists s \in \mathcal{X} \setminus \mathcal{X}_0)(\forall r \in \mathcal{X}_0) r \perp s$. Thus, by elementarity, $\mathcal{M} \models (\exists s \in \mathcal{X} \setminus \mathcal{X}_0)(\forall r \in \mathcal{X}_0) r \perp s$. As $\mathcal{X}_0$ was arbitrary, we in fact have

$$\mathcal{M} \models (\forall \text{ countable } \mathcal{X}_0 \subseteq \mathcal{X})(\exists s \in \mathcal{X} \setminus \mathcal{X}_0)(\forall r \in \mathcal{X}_0) r \perp s.$$

Elementarity again will give us that

$$H_\theta \models (\forall \text{ countable } \mathcal{X}_0 \subseteq \mathcal{X})(\exists s \in \mathcal{X} \setminus \mathcal{X}_0)(\forall r \in \mathcal{X}_0) r \perp s.$$

However, if $\mathcal{X}_0 \subseteq \mathcal{X}$ is a maximal family of pairwise incompatible elements of $\mathcal{X}$, then by the c.c.c. $\mathcal{X}_0$ is countable, and trivially each condition in $\mathcal{X} \setminus \mathcal{X}_0$ is compatible with some condition in $\mathcal{X}_0$.

By this contradiction there is an $s \in \mathcal{X} \cap \mathcal{M}$ which is compatible with $\bar{r}$. As $s \in \mathcal{X}$, then by elementarity we have that $\mathcal{M} \models (\exists \gamma > \beta)(\gamma \in C(s))$, and therefore there is a $\gamma \in \omega_1 \cap \mathcal{M} = \delta$ with $\gamma \in C(s)$ and $\gamma > \beta$. Then, obviously $C(s) \cap (\beta, \delta] \neq \emptyset$.

Let $t \in \mathcal{P}$ be a common extension of $\bar{r}$ and s. By $\mathcal{P}$-monotonicity it follows that $C(t) \cap (\beta, \delta] \neq \emptyset$. However, this contradicts our choice of $\bar{r}$. $\square$

We then immediately have that $C(q) \cap S \neq \emptyset$. It then follows that $\mathcal{P}$ is stationary set preserving. $\square$

Corollary 4. $\mathrm{mm} \leq \mathrm{m}$.

The proof of the above Theorem is useful for pedagogical reasons that will become apparent in a future section. Note, however, that the following statement, which immediately implies the above Lemma, is also true:

Proposition. *Let $\mathcal{P}$ is a given c.c.c.-poset, and let $\langle C(p) : p \in \mathcal{P} \rangle$ be a $\mathcal{P}$-club. Then the set*

$$\{p \in \mathcal{P} : C(p) \text{ is a club in } \omega_1\}$$

is dense in $\mathcal{P}$.

Theorem 5. $\mathrm{mm} \leq \omega_2$.

Proof. Consider the set $\mathcal{P}$ of all countable partial functions $\omega_1 \to \omega_2$, and order $\mathcal{P}$ by inverse-inclusion. Note that $\mathcal{P}$ is σ-closed, as any countable union of countable sets is again countable. A fact to be proved in the next section shows that all σ-closed partial orders are stationary set preserving.

For each $\alpha < \omega_2$ define

$$D_\alpha = \{p \in \mathcal{P} : \alpha \in \mathrm{rng}(p)\}.$$

It is trivial to check that each D_α is dense-open in $\mathcal{P}$.

Suppose that $\mathcal{G}$ is a filter in $\mathcal{P}$ which meets each D_α. As $\mathcal{G}$ is a filter, then $g = \bigcup \mathcal{G}$ is a function. As $\mathcal{G} \cap D_\alpha \neq \emptyset$, then $\alpha \in \mathrm{rng}(g)$, and thus $\omega_2 = \mathrm{rng}(g)$. But this is impossible, since $\mathrm{dom}(g) \subseteq \omega_1 < \omega_2$.

It then follows that there is no filter in $\mathcal{P}$ which meets each of the D_α, and thus $\mathfrak{m}(\mathcal{P}) \leq \omega_2$. $\qquad\qquad\square$

1.5   Proper and semi-proper posets

Definition. A partial order $\mathcal{P}$ is called σ-*closed* if every decreasing sequence $\{p_n\}_{n \in \omega}$ in $\mathcal{P}$ has a lower bound in $\mathcal{P}$.

Definition. A partial order $\mathcal{P}$ is said to satisfy the *countable chain condition*, or simply is ccc, if every family of pairwise incompatible elements of $\mathcal{P}$ is countable.

The notion of properness will be a generalisation of both the c.c.c. and σ-closedness. Before we given the definition, let us look at a comparison of σ-closed partial orders, and stationary-set preserving partial orders.

Theorem 6. *Every σ-closed partial order $\mathcal{P}$ is stationary set preserving.*

Proof. Let $\mathcal{P}$ be a σ-closed partial order, let $\langle C(p) : p \in \mathcal{P} \rangle$ be a $\mathcal{P}$-club, and let $S \subseteq \omega_1$ be stationary. Let $\bar{p} \in \mathcal{P}$ be a given condition. Take a countably elementary submodel $\mathcal{M} \prec \langle H_\theta, \in \rangle$ for some large enough θ containing $\mathcal{P}, \langle C(p) : p \in \mathcal{P} \rangle, S, \bar{p}$, and such that $\delta = \omega_1 \cap \mathcal{M} \in S$. Let $\{\mathcal{X}_n : n \in \omega\}$ be an enumeration of all subsets of $\mathcal{P}$ that belong to $\mathcal{M}$.

Using elementarity, and starting from $p_0 = \bar{p}$, find a decreasing sequence $\{p_n\}_{n \in \omega}$ of elements of $\mathcal{P} \cap \mathcal{M}$ such that for each $n \in \omega$,

$$\begin{cases} p_{n+1} \in \mathcal{X}_n, & \text{if there is a } q \in \mathcal{X}_n \text{ with } q \leq p_n, \text{ and} \\ p_{n+1} = p_n, & \text{otherwise.} \end{cases}$$

By σ-closedness, let $q \in \mathcal{P}$ be a lower bound of $\{p_n : n \in \omega\}$. Using $\mathcal{P}$-unboundedness, we may assume that $C(q) \not\subseteq \delta$.

Claim 6.1. $\delta \in C(q)$.

Proof of claim. If not, then by $\mathcal{P}$-closedness there is a $\beta < \delta$ and a $\bar{r} \leq q$ such that $C(s) \cap (\beta, \delta] = \emptyset$ for each $s \leq \bar{r}$. As $\beta \in \mathcal{M}$, then the set $\mathcal{X} = \{r \leq \bar{p} : C(r) \nsubseteq \beta + 1\}$ is in $\mathcal{M}$. Then $\mathcal{X} = \mathcal{X}_n$ for some $n \in \omega$.

It follows that at stage n in the above recursion, it must have been that p_{n+1} was chosen to be an element of $\mathcal{X}_n$. Since $p_{n+1} \in \mathcal{M}$, by definition of $\mathcal{X}$ and elementarity it follows $\mathcal{M} \models (\exists \gamma > \beta)\gamma \in C(p_{n+1})$, and therefore there is a $\gamma \in \omega_1 \cap \mathcal{M} = \delta$ with $\gamma \in C(p_{n+1})$ and $\gamma > \beta$. Then we trivially have that $C(p_{n+1}) \cap (\beta, \delta] \neq \emptyset$. As $\bar{r} \leq q \leq p_{n+1}$, by $\mathcal{P}$-monotonicity, this clearly contradicts the choice of $\bar{r}$. □

As $\delta \in S$, it immediately follows that

$$\{p \in \mathcal{P} : C(p) \cap S \neq \emptyset\}$$

is dense in $\mathcal{P}$, and so $\mathcal{P}$ is stationary set preserving. □

Note that in the proofs that every partial order with the c.c.c. is stationary set preserving, and that every σ-closed partial order is stationary set preserving, we have the following general scheme:

Step 1: Start with a partial order $\mathcal{P}$, a $\mathcal{P}$-club $\langle C(p) : p \in \mathcal{P}\rangle$, a stationary set $S \subseteq \omega_1$ and a given condition $\bar{p}$.

Step 2: Take a countable elementary submodel $\mathcal{M} \prec \langle H_\theta, \in \rangle$ for some large enough θ which contains each of the objects from Step 1.

Step 3: Extend the given condition $\bar{p}$ to some q. We then assume that q does not have some good property (in our cases, we assume that $\delta = \omega_1 \cap \mathcal{M}$ is not in $C(q)$). This leads us to a further extension $\bar{r}$ of q, each further extension of which is "bad".

Step 4: We define some subset $\mathcal{X}$ of $\mathcal{P}$ in $\mathcal{M}$ which contains $\bar{r}$, and which has some "nice" property of $\bar{r}$.

Step 5: Show that $\bar{r}$ is "reflected" in $\mathcal{X} \cap \mathcal{M}$; i.e., show that $\bar{r}$ is compatible with some element s of $\mathcal{X} \cap \mathcal{M}$.

Step 6: Taking a common extension t of $\bar{r}$ and s, arrive at a contradiction.

The most important step in the above is Step 5. Upon further inspection, we may see that the actual definition of the set $\mathcal{X}$ was unimportant. All that was required of $\mathcal{X}$ was that it was a nonempty subset of $\mathcal{P}$ in $\mathcal{M}$ containing the condition $\bar{r}$. Once this was accomplished, we are guaranteed that $\bar{r}$ will "reflect" in $\mathcal{X} \cap \mathcal{M}$. It is this notion of "reflecting" that will be extremely important to the concept of properness.

Definition. Let $\mathcal{P}$ be a partial order, and $\mathcal{M}$ a countable elementary sub-model of $\langle H_\theta, \in \rangle$ for some large enough θ with $\mathcal{P} \in \mathcal{M}$. We say that $q \in \mathcal{P}$ is $(\mathcal{M}, \mathcal{P})$-*generic* (or, simply $\mathcal{M}$-*generic*) if for every $r \le q$ and $\mathcal{X} \subseteq \mathcal{P}$ such that $\mathcal{X} \in \mathcal{M}$ and $r \in \mathcal{X}$ there is $\bar{r} \in \mathcal{X} \cap \mathcal{M}$ such that r and $\bar{r}$ are compatible.

The notion of properness will simply state that there are always many $(\mathcal{M}, \mathcal{P})$-generic conditions:

Definition. A partial order $\mathcal{P}$ is called *proper* if for each countable ele-mentary submodel $\mathcal{M} \prec \langle H_\theta, \in, \rangle$ for large enough θ the set and for every $p \in \mathcal{P} \cap M$ there is $q \le p$ such that q is $(\mathcal{M}, \mathcal{P})$-generic.

The proofs given above readily give us the following:

Corollary 7.   *1. Every c.c.c. partial order is proper.*

2. Every σ-closed partial order in proper.

Corollary 8. *Every proper partial order is stationary set preserving.*

Exercise. Give an example of a proper poset $\mathcal{P}$ that is neither c.c.c. nor is σ-closed. Can you find such a poset $\mathcal{P}$ to be of cardinality $\aleph_1$ or $\aleph_2$?

Exercise. Let T be a tree of height ω_1. Under which conditions is T as a forcing notion proper?

There is another less restrictive condition on a poset $\mathcal{P}$ that is still stronger than the condition of preserving all stationary subsets of ω_1.

Definition. Let $\mathcal{P}$ be a partial order, and $\mathcal{M}$ a countable elementary sub-model of $\langle H_\theta, \in \rangle$ for some large enough θ with $\mathcal{P} \in \mathcal{M}$. We say that $q \in \mathcal{P}$ is $(\mathcal{M}, \mathcal{P})$-*semi-generic* (or, simply $\mathcal{M}$-*semi-generic*) if for every $r \le q$ and every partial function $f : \mathcal{P} \to \omega_1$ such that $f \in \mathcal{M}$ and $r \in \mathrm{dom}(f)$ there exist $\alpha \in \mathcal{M} \cap \omega_1$ and $\bar{r} \in f^{-1}(\alpha)$ such that r and $\bar{r}$ are compatible.

Definition. A partial order $\mathcal{P}$ is called *semi-proper* if for each countable elementary submodel $\mathcal{M} \prec \langle H_\theta, \in, \rangle$ for large enough θ the set and for every $p \in \mathcal{P} \cap M$ there is $q \le p$ such that q is $(\mathcal{M}, \mathcal{P})$-semi-generic.

Clearly, every proper poset is semi-proper.

Exercise. Give an example of a semi-proper poset that is not proper. What is the minimal cardinal κ for which you can find a semi-proper poset that does not preserve stationary subsets of $[\kappa]^{\aleph_0}$?

Exercise. Let T be a tree of height ω_1. Under which conditions is T as a forcing notion semi-proper?

Theorem 9. *Every semi-proper posets preserves all stationary subsets of* ω_1.

Proof. Let $(C_p : p \in \mathcal{P})$ be a given $\mathcal{P}$-club, Let E be a given stationary subset of ω_1 and let $\bar{p}$ be a given condition of $\mathcal{P}$. WE need to find $q \leq \bar{p}$ such that $C(q) \cap E \neq \emptyset$. Choose a countable elementary submodel M of some large enough structure of the form $(H_\theta, \in)$ such that $\delta = M \cap \omega_1$ belongs to E and such that M contains all the relevant objects such as $\mathcal{P}$, $(C_p : p \in \mathcal{P})$, and E. Choose $p \leq \bar{p}$ such that p is $(M, \mathcal{P})$-semi-generic and then choose $q \leq p$ such that $C(q) \setminus \delta \neq \emptyset$.

Claim 9.1. $\delta \in C(q)$.

Proof. Otherwise, we can find $r \leq q$ and $\gamma < \delta$ such that $C(s) \cap (\gamma, \delta] \neq \emptyset$ for all $s \leq r$. Let

$$\mathcal{X} = \{x \in \mathcal{P} : x \leq \bar{p} \text{ and } C(x) \setminus \gamma \neq \emptyset\}.$$

Clearly $\mathcal{X} \in M$ and $q \in \mathcal{X}$. Define $f : \mathcal{X} \to \omega_1$ by letting

$$f(x) = \min(C(x) \setminus \gamma).$$

Clearly $f \in M$. Since p is an $(M, \mathcal{P})$-semi-generic condition there exist $\alpha < \delta = M \cap \omega_1$ and $\bar{r} \in f^{-1}(\alpha)$ such that $\bar{r}$ and r are compatible in $\mathcal{P}$. Let s be a common extension of $\bar{r}$ and r. Then $\alpha \in C(s)$. So in particular, $C(s) \cap (\gamma, \delta] \neq \emptyset$, a contradiction.   □

It is clear that the Claim finishes the proof.   □

The reader might wonder if there is any difference between the notion of stationary set preserving and the notion of semi-proper. This is indeed the right question as its resolution was directly responsible for some of the major discoveries in set theory in the 1980's. The reader is invited to complete the following exercises that test the same.

Exercise. Show that there is a stationary set preserving poset $\mathcal{P}$ of size $\aleph_2$ that collapses $\aleph_2$ to $\aleph_1$.

Note that such poset $\mathcal{P}$ necessarily preserves all cardinals $> \aleph_2$.

Exercise. Show that if a semi-proper poset $\mathcal{P}$ collapses $\aleph_2$ to $\aleph_1$ then it necessarily collapses the (ground model) continuum to $\aleph_1$. So, in particular, if $\mathcal{P}$ has size at most continuum then $\mathcal{P}$ forces CH (and, in fact, $\diamondsuit$).

Exercise. Show that for every positive integer n there is a stationary set preserving poset $\mathcal{P}$ of size $\aleph_n$ that collapses $\aleph_n$ to $\aleph_1$.

Problem 1.5.1.   (1) Is there a stationary set preserving poset of size $\aleph_{\omega+1}$ that collapses $\aleph_{\omega+1}$ to $\aleph_1$?

(2) In general, for which ordinal α we can find a poset $\mathcal{P}$ of cardinality $\aleph_\alpha$ which collapses $\aleph_\alpha$ to $\aleph_1$.

(3) More generally, for which ordinals α there is a poset $\mathcal{P}$ that collapses $\aleph_\alpha$ to $\aleph_1$, but preserves all cardinals $> \aleph_\alpha$.

Chapter 2

Coding Sets by the Real Numbers

2.1   Almost-disjoint coding

Definition. A family $\mathcal{A}$ of infinite subsets of $\mathbb{N}$ is called *almost disjoint (a.d.)* if $A \cap B \in \textit{Fin}$ for any distinct $A, B \in \mathcal{A}$.

Proposition (W. Sierpinski, 1930's)**.** *There is an a.d. family $\mathcal{A}$ of size $\mathfrak{c}$.*

Proof. We will instead construct an almost disjoint family of infinite subsets of $\mathbb{Q}$. This is clearly sufficient, and any bijection $\mathbb{Q} \to \mathbb{N}$ will translate such a family into an a.d. family of infinite subsets of $\mathbb{N}$.

For each real number x fix a nonrepeating sequence $\{q_n^{(x)}\}_{n<\omega}$ of rationals converging to x, and let $A_x = \{q_n^{(x)} : n \in \omega\}$. Clearly each A_x is an infinite subset of $\mathbb{Q}$, and for any $x \neq y$ it must be that $A_x \cap A_y$ is finite. $\square$

Definition. Fix an a.d. family $\mathcal{A}$. Given any $\mathcal{B} \subseteq \mathcal{A}$, we say that $X_{\mathcal{B}} \subseteq \mathbb{N}$ is a *code for $\mathcal{B}$ in $\mathcal{A}$* iff

$$(\forall A \in \mathcal{A})\ A \in \mathcal{B} \leftrightarrow A \cap X_{\mathcal{B}} \in \textit{Fin}.$$

Theorem 10. *Given an a.d. family $\mathcal{A}$ with $|\mathcal{A}| < \mathfrak{p}$, and a $\mathcal{B} \subseteq \mathcal{A}$, there is a code for $\mathcal{B}$ in $\mathcal{A}$.*

Proof. Let $\mathcal{P}$ be the collection of all pairs $p = \langle X_p, \mathcal{F}_p \rangle$ where

(1) $X_p \in \textit{Fin}$, and

(2) $\mathcal{F}_p$ is a finite subset of $\mathcal{B}$.

We order $\mathcal{P}$ by defining $p \leq q$ iff

13

(3) $X_p \supseteq X_q$, and

(4) $\max(X_q) < \min(X_p \setminus X_q)$ (where $\max(\emptyset) = -\infty$ and $\min(\emptyset) = \infty$), and

(5) $\mathcal{F}_p \supseteq \mathcal{F}_q$, and

(6) $(X_p \setminus X_q) \cap \bigcup \mathcal{F}_q = \emptyset$.

Claim 10.1. $\langle \mathcal{P}, \leq \rangle$ *is a partial order.*

Proof of claim. Reflexivity is trivial, while antisymmetry follows easily from (3) and (5).

The only difficulty in showing transitivity is (6), which follows once seeing that $X_p \supseteq X_q \supseteq X_r$ implies $X_p \setminus X_r = (X_p \setminus X_q) \cup (X_q \setminus X_r)$. □

Claim 10.2. $\langle \mathcal{P}, \leq \rangle$ *is σ-centered.*

Proof of claim. Suppose that $A \subseteq \mathcal{P}$ is uncountable. It easily follows that there are distinct $p, q \in A$ such that $X_p = X_q$. Consider $r = \langle X_p, \mathcal{F}_p \cup \mathcal{F}_q \rangle$. It is easy to show that $r \leq p, q$, and therefore p and q are compatible. Thus, any antichain in $\langle \mathcal{P}, \leq \rangle$ is countable. □

Claim 10.3. *For each $A \in \mathcal{A} \setminus \mathcal{B}$, and any $k \in \mathbb{N}$ the set*

$$\mathcal{D}_{A,k} = \{p \in \mathcal{P} : (X_p \cap A) \setminus \{0, \ldots, k\} \neq \emptyset\}$$

is dense-open in $\langle \mathcal{P}, \leq \rangle$.

Proof of claim. Suppose that $q \in \mathcal{P}$ but $q \notin \mathcal{D}_{A,k}$.

Note that as $\mathcal{A}$ is a.d. and $\mathcal{F}_q \subseteq \mathcal{A}$ is finite, we have that $A \setminus \bigcup \mathcal{F}_q$ is infinite. Therefore there is an $\ell \in A \setminus \bigcup \mathcal{F}_q$ with $\ell > k, \max(X_q)$. Define $p = \langle X_q \cup \{\ell\}, \mathcal{F}_q \rangle$. It is easy to show that $p \in \mathcal{D}_{A,k}$ and $p \leq q$.

Given $p \leq q \in \mathcal{P}$ with $q \in \mathcal{D}_{A,k}$, we trivially have

$$(X_p \cap A) \setminus \{0, \ldots, k\} \supseteq (X_q \cap A) \setminus \{0, \ldots, k\} \neq \emptyset,$$

and therefore $p \in \mathcal{D}_{A,k}$. □

Claim 10.4. *For each $A \in \mathcal{B}$ the set*

$$\mathcal{E}_A = \{p \in \mathcal{P} : A \in \mathcal{F}_p\}$$

is dense-open in $\langle \mathcal{P}, \leq \rangle$.

Proof of claim. Suppose $q \in \mathcal{P}$ but $q \notin \mathcal{E}_A$. Define $p = \langle X_q, \mathcal{F}_q \cup \{A\} \rangle$. Clearly $p \in \mathcal{E}_A$, and $p \leq q$.

Suppose that $p \leq q \in \mathcal{P}$ and $q \in \mathcal{E}_A$. Then we clearly have $A \in \mathcal{F}_q \subseteq \mathcal{F}_p$, and therefore $p \in \mathcal{E}_A$. □

Note that in Claims 10.3 and 10.4 we have listed at most $\max\{|\mathcal{A}|, \aleph_0\} < \mathfrak{m}$ distinct dense-open sets in $\langle \mathcal{P}, \leq \rangle$, and therefore there is a filter $\mathcal{G}$ in $\mathcal{P}$ which meets each of these. Define $X_{\mathcal{B}} = \bigcup\{X_p : p \in \mathcal{G}\}$.

Claim 10.5. $X_{\mathcal{B}}$ *is a code for* $\mathcal{B}$ *in* $\mathcal{A}$.

Proof of claim. Let $A \in \mathcal{A}$.

- Suppose that $A \notin \mathcal{B}$. We will show that for all $k \in \mathbb{N}$ the set $(A \cap X_{\mathcal{B}}) \setminus \{0, 1, \ldots, k\}$ is nonempty. As $\mathcal{G} \cap D_{A,k} \neq \emptyset$, there is a $p \in \mathcal{G} \cap D_{A,k}$. Trivially as $X_p \subseteq X_{\mathcal{B}}$ we have

$$(X_{\mathcal{B}} \cap A) \setminus \{0, 1, \ldots, k\} \supseteq (X_p \cap A) \setminus \{0, 1, \ldots, k\} \neq \emptyset.$$

- Suppose that $A \in \mathcal{B}$. As $\mathcal{G} \cap \mathcal{E}_A \neq \emptyset$ there is a $p \in \mathcal{G}$ with $A \in \mathcal{F}_p$. We will show that $A \cap X_{\mathcal{B}} \subseteq X_p$, which will clearly show that $A \cap X_{\mathcal{B}}$ is finite.

  Given any $\ell \in A \cap X_{\mathcal{B}}$, there is a $q \in \mathcal{G}$ such that $\ell \in X_q$. As $p, q \in \mathcal{G}$, there is an $r \in \mathcal{G}$ with $r \leq p, q$. As $r \leq p$ and $A \in \mathcal{F}_p$ we have $(X_r \setminus X_p) \cap A = \emptyset$. However, as $r \leq q$ we have $\ell \in X_r$, and thus $\ell \in X_r \cap A$. It then must be that $\ell \in X_p$. $\square$

This then completes the proof. $\square$

Corollary. *For any* $\kappa < \mathfrak{p}$ *we have* $2^\kappa \leq \mathfrak{c}$.

Proof. As $\mathfrak{p} \leq \mathfrak{c}$, there is an a.d. family $\mathcal{A}$ of size κ for any $\kappa < \mathfrak{p}$. Using the above theorem, for each $\mathcal{B} \subseteq \mathcal{A}$, there is a code $X_{\mathcal{B}}$ for $\mathcal{B}$ in $\mathcal{A}$. Trivially, for $\mathcal{B} \neq \mathcal{D} \subseteq \mathcal{A}$ we have $X_{\mathcal{B}} \neq X_{\mathcal{D}}$, and therefore the mapping $\mathcal{B} \mapsto X_{\mathcal{B}}$ is an injection $\mathcal{P}(\mathcal{A}) \to \mathcal{P}(\mathbb{N})$. $\square$

2.2 Coding families of unordered pairs of ordinals

Theorem 11. *Suppose* $S \subseteq [\mathfrak{p}]^2$. *Then there is a sequence* $\{A_\alpha\}_{\alpha < \mathfrak{m}}$ *of infinite subsets of* $\mathbb{N}$ *such that for any* $\{\alpha, \beta\} \in [\mathfrak{p}]^2$

$$\{\alpha, \beta\} \in S \quad \Leftrightarrow \quad A_\alpha \cap A_\beta \in Fin.$$

Proof. We will inductively construct the sequence so that for any $\gamma < \mathfrak{m}$ it satisfies:

$(1)_\gamma$ $\{\alpha, \beta\} \in S \Leftrightarrow |A_\alpha \cap A_\beta| < \aleph_0$ for all $\{\alpha, \beta\} \in [\gamma]^2$, and

$(2)_\gamma$ $A_\alpha \setminus \bigcup_{\beta \in \Gamma} A_\beta$ is infinite for all $\alpha < \gamma$ and all finite $\Gamma \subseteq \gamma \setminus \{\alpha\}$, and

$(3)_\gamma$ $\mathbb{N} \setminus \bigcup_{\beta \in \Gamma} A_\beta$ is infinite for all finite $\Gamma \subseteq \gamma$.

It is clear that $(2)_\gamma$ guarantees that the A_α will be infinite. It is also clear
that $(3)_\gamma$ is an immediate consequence of $(2)_\gamma$ when $\gamma \geq \aleph_0$. We will
then start our induction at $\gamma = \omega$, and show that we can find a sequence
$\{A_n\}_{n<\omega}$ satisfying $(1)_\omega$ and $(2)_\omega$. To start, let $\{p_n\}_{n<\omega}$ enumerate the
prime numbers. For each $n < \omega$ define

$$A_n = \{p_n^i : i > 0\} \cup \bigcup \{\{p_n^i p_m^j : i,j > 0\} : m \in \omega \setminus \{n\}, \{m,n\} \notin S\}.$$

It is easy to check that $\{A_n\}_{n<\omega}$ satisfies $(1)_\omega$ and $(2)_\omega$.

Suppose that for some $\omega \leq \gamma < \mathfrak{p}$ a sequence $\{A_\alpha\}_{\alpha<\gamma}$ subsets of $\mathbb{N}$ has
been chosen to satisfy $(1)_\gamma$ and $(2)_\gamma$. Define $\mathcal{P}$ to be the collection of all
pairs $p = \langle X_p, \mathcal{F}_p \rangle$ where

(4) $X_p \in Fin$, and

(5) $\mathcal{F}_p$ is a finite subset of $\{\alpha < \gamma : \{\alpha,\gamma\} \in S\}$.

We order $\mathcal{P}$ by defining $p \leq q$ iff

(6) $X_p \supseteq X_q$, and

(7) $\max(X_q) < \min(X_p \setminus X_q)$, and

(8) $\mathcal{F}_p \supseteq \mathcal{F}_q$, and

(9) $(X_p \setminus X_q) \cap \bigcup_{\beta \in \mathcal{F}_q} A_\beta = \emptyset$.

Claim 11.1. $\langle \mathcal{P}, \leq \rangle$ *is a σ-centered partial order.*

Proof of claim. Similar to the proofs of Claims 10.1 and 10.2. $\square$

The following claim will be useful in extending conditions in $\mathcal{P}$:

Claim 11.2. *Suppose that $q \in \mathcal{P}$ and $\ell \in \mathbb{N}$ are such that $\ell \notin \bigcup_{\beta \in \mathcal{F}_q} A_\beta$
and $\ell > \max(X_q)$. Then defining $p = \langle X_q \cup \{\ell\}, \mathcal{F}_q \rangle$ we have $p \leq q$.*

Proof of claim. This is trivial as $X_p \setminus X_q = \{\ell\}$, and by choice we have
$\ell > \max(X_q)$ and also $\ell \notin \bigcup_{\beta \in \mathcal{F}_q} A_\beta$. $\square$

Claim 11.3. *The following subsets of $\mathcal{P}$ are dense-open in $\langle \mathcal{P}, \leq \rangle$:*

(i) *For each $\alpha < \gamma$ with $\{\alpha,\gamma\} \in S$ the set*

$$\mathcal{D}_\alpha^{(1)} = \{p \in \mathcal{P} : \alpha \in \mathcal{F}_p\}.$$

(ii) *For each $\alpha < \gamma$ with $\{\alpha,\gamma\} \notin S$ and each $k \in \mathbb{N}$ the set*

$$\mathcal{D}_{\alpha,k}^{(2)} = \{p \in \mathcal{P} : (X_p \cap A_\alpha) \setminus \{0,\ldots,k\} \neq \emptyset\}.$$

(iii) *For each finite* $\Gamma \subseteq \gamma$ *and each* $k \in \mathbb{N}$ *the set*

$$\mathcal{D}^{(3)}_{\Gamma,k} = \{p \in \mathcal{P} : (X_p \setminus \bigcup_{\beta \in \Gamma} A_\beta) \setminus \{0, \dots, k\} \neq \emptyset\}.$$

(iv) *For each* $\alpha < \gamma$, *each finite* $\Gamma \subseteq \gamma \setminus \{\alpha\}$ *and each* $k \in \mathbb{N}$ *the set*

$$\mathcal{D}^{(4)}_{\alpha,\Gamma,k} = \{p \in \mathcal{P} : (A_\alpha \setminus (X_p \cup \bigcup_{\beta \in \Gamma} A_\beta)) \cap \{k, \dots, \max(X_p)\} \neq \emptyset\}.$$

Proof of claim. (i) Let $\alpha < \gamma$ be such that $\{\alpha, \gamma\} \in S$. Suppose that $q \in \mathcal{P}$ but $q \notin \mathcal{D}^{(1)}_\alpha$. Define $p = \langle X_q, \mathcal{F}_q \cup \{\alpha\}\rangle$. Trivially $p \in \mathcal{D}^{(1)}_\alpha$, and also $p \leq q$.

Suppose that $p \leq q \in \mathcal{P}$ and $q \in \mathcal{D}^{(1)}_\alpha$. Clearly $\mathcal{F}_p \supseteq \mathcal{F}_q \ni \alpha$, and therefore $p \in \mathcal{D}^{(1)}_\alpha$.

(ii) Let $\alpha < \gamma$ be such that $\{\alpha, \gamma\} \notin S$ and let $k \in \mathbb{N}$. Suppose that $q \in \mathcal{P}$ but $q \notin \mathcal{D}^{(2)}_{\alpha,k}$. As $A_\alpha \setminus \bigcup_{\beta \in \mathcal{F}_q} A_\beta$ is infinite, we may choose some $\ell \in A_\alpha \setminus \bigcup_{\beta \in \mathcal{F}_q} A_\beta$ with $\ell > k, \max(X_q)$. Define $p = \langle X_q \cup \{\ell\}, \mathcal{F}_q\rangle$. Clearly, $\ell \in X_p \cap A_\alpha$ and $\ell > k$, so $p \in \mathcal{D}^{(2)}_{\alpha,k}$. By Claim 11.2 we also have $p \leq q$.

Suppose that $p \leq q \in \mathcal{P}$ and $q \in \mathcal{D}^{(2)}_{\alpha,k}$. As $X_p \supseteq X_q$, we clearly have $(X_p \cap A_\alpha) \setminus \{0, \dots, k\} \supseteq (X_q \cap A_\alpha) \setminus \{0, \dots, k\} \neq \emptyset$. Thus $p \in \mathcal{D}^{(2)}_{\alpha,k}$.

(iii) Let $\Gamma \subseteq \gamma$ be finite, and let $k \in \mathbb{N}$. Suppose that $q \in \mathcal{P}$ but $q \notin \mathcal{D}^{(3)}_{\Gamma,k}$. By $(3)_\gamma$ we may choose some $\ell \in \mathbb{N} \setminus \bigcup_{\beta \in \Gamma \cup \mathcal{F}_q} A_\beta$ with $\ell > k, \max(X_q)$. Define $p = \langle X_q \cup \{\ell\}, \mathcal{F}_q\rangle$. By choice we have $\ell \in X_p \setminus \bigcup_{\beta \in \Gamma} A_\beta$. Thus, as $\ell > k$, we have $p \in \mathcal{D}^{(3)}_{\Gamma,k}$. By Claim 11.2 we also have $p \leq q$.

Suppose that $p \leq q \in \mathcal{P}$ and $q \in \mathcal{D}^{(3)}_{\Gamma,k}$. As $X_p \supseteq X_q$ we trivially have

$$(X_p \setminus \bigcup_{\beta \in \Gamma} A_\beta) \setminus \{0, \dots, k\} \supseteq (X_q \setminus \bigcup_{\beta \in \Gamma} A_\beta) \setminus \{0, \dots, k\} \neq \emptyset.$$

Thus $p \in \mathcal{D}^{(3)}_{\Gamma,k}$.

(iv) Let $\alpha < \gamma$, $\Gamma \subseteq \gamma \setminus \{\alpha\}$ be finite, and $k \in \mathbb{N}$. Suppose that $q \in \mathcal{P}$ but $q \notin \mathcal{D}^{(4)}_{\alpha,\Gamma,k}$. As $A_\alpha \setminus \bigcup_{\beta \in \Gamma} A_\beta$ is infinite and X_q is finite, then $A_\alpha \setminus (X_q \cup \bigcup_{\beta \in \Gamma} A_\beta)$ is also infinte, so we may choose some $\ell' \in A_\alpha \setminus (X_q \cup \bigcup_{\beta \in \Gamma} A_\beta)$ with $\ell' > k$. By $(3)_\gamma$ we may choose some $\ell \in \mathbb{N} \setminus \bigcup_{\beta \in \mathcal{F}_q} A_\beta$ with $\ell > \ell', \max(X_q)$. Define $p = \langle X_q \cup \{\ell\}, \mathcal{F}_q\rangle$. Clearly $\ell' \notin X_p$ and so by choice we have $\ell' \in A_\alpha \setminus (X_p \cup \bigcup_{\beta \in \Gamma} A_\beta)$.

Also by choice we have $k \leq \ell' < \ell = \max(X_p)$, and so $p \in \mathcal{D}^{(4)}_{\alpha,\Gamma,k}$. By Claim 11.2 we also have $p \leq q$.

Suppose that $p \leq q \in \mathcal{P}$ and $q \in \mathcal{D}^{(4)}_{\alpha,\Gamma,k}$. Pick $i \in (A_\alpha \setminus (X_q \cup \bigcup_{\beta \in \Gamma} A_\beta)) \cap \{k, \ldots, \max(X_q)\}$. As $\max(X_q) \leq \max(X_p)$ we have $i \in \{k, \ldots, \max(X_p)\}$. As $i \leq \max(X_q) < \min(X_p \setminus X_q)$, and $i \notin X_q$ it follows that $i \notin X_p$, and therefore $i \in A_\alpha \setminus (X_p \cup \bigcup_{\beta \in \Gamma} A_\beta)$. Therefore $p \in \mathcal{D}^{(4)}_{\alpha,\Gamma,k}$. $\hfill\square$

Note that in Claim 11.3 we have listed at most $|\gamma| < \mathfrak{p}$ distinct dense-open subsets of $\mathcal{P}$, and so there is a filter $\mathcal{G}$ in $\langle \mathcal{P}, \leq \rangle$ meeting each of these. Define $A_\gamma = \bigcup_{p \in \mathcal{G}} X_p$. To show that $\{A_\alpha\}_{\alpha < \gamma + 1}$ satisfies $(1)_{\gamma+1}$ and $(2)_{\gamma+1}$ it clearly suffices to prove the following:

Claim 11.4.   *(i) $A_\gamma \cap A_\alpha$ is finite for all $\alpha < \gamma$ with $\{\alpha, \gamma\} \in S$.*

(ii) $A_\gamma \cap A_\alpha$ is infinite for all $\alpha < \gamma$ with $\{\alpha, \gamma\} \notin S$.

(iii) $A_\gamma \setminus \bigcup_{\beta \in \Gamma} A_\beta$ is infinite for all finite $\Gamma \subseteq \gamma$.

(iv) $A_\alpha \setminus (A_\gamma \cup \bigcup_{\beta \in \Gamma} A_\beta)$ is infinite for all $\alpha < \gamma$ and all finite $\Gamma \subseteq \gamma \setminus \{\alpha\}$.

Proof.   (i) Let $\alpha < \gamma$ be such that $\{\alpha, \gamma\} \in S$. Then pick some $p \in \mathcal{G} \cap \mathcal{D}^{(1)}_\alpha$. It suffices to show that $A_\alpha \cap A_\gamma \subseteq X_p$, as X_p is finite. Suppose that $i \in A_\alpha \cap A_\gamma$. Then there is a $q \in \mathcal{G}$ with $i \in X_q$. As $\mathcal{G}$ is a filter, there is an $r \in \mathcal{G}$ with $r \leq p, q$. As $r \leq q$ we have $X_r \supseteq X_q \ni i$, and therefore $i \in X_r \cap A_\alpha$. As $r \leq p$ we have $(X_r \setminus X_p) \cap A_\alpha = \emptyset$, and therefore it follows that $i \in X_p$. Thus $A_\alpha \cap A_\gamma \subseteq X_p$, as desired.

(ii) Let $\alpha < \gamma$ be such that $\{\alpha, \gamma\} \notin S$. To show $A_\gamma \cap A_\alpha$ is infinite, it suffices to show that for each $k \in \mathbb{N}$ the set $(A_\gamma \cap A_\alpha) \setminus \{0, \ldots, k\}$ is nonempty. Given $k \in \mathbb{N}$, pick some $p \in \mathcal{G} \cap \mathcal{D}^{(2)}_{\alpha,k}$. As $A_\gamma \supseteq X_p$ we trivially have

$$(A_\gamma \cap A_\alpha) \setminus \{0, \ldots, k\} \supseteq (X_p \cap A_\alpha) \setminus \{0, \ldots, k\} \neq \emptyset.$$

(iii) Let $\Gamma \subseteq \gamma$ be finite. To show $A_\gamma \setminus \bigcup_{\beta \in \Gamma} A_\beta$ is infinite, it suffices to show that for each $k \in \mathbb{N}$ the set $(A_\gamma \setminus \bigcup_{\beta \in \Gamma} A_\beta) \setminus \{0, \ldots, k\}$ is nonempty. Given $k \in \mathbb{N}$ pick some $p \in \mathcal{G} \cap \mathcal{D}^{(3)}_{\Gamma,k}$. As $A_\gamma \supseteq X_p$ we trivially have

$$\left(A_\gamma \setminus \bigcup_{\beta \in \Gamma} A_\beta\right) \setminus \{0, \ldots, k\} \supseteq \left(X_p \setminus \bigcup_{\beta \in \Gamma} A_\beta\right) \setminus \{0, \ldots, k\} \neq \emptyset.$$

(iv) Let $\alpha < \gamma$, and $\Gamma \subseteq \gamma \backslash \{\alpha\}$ be finite. To show that $A_\alpha \backslash (A_\gamma \cup \bigcup_{\beta \in \Gamma} A_\beta)$ is infinite, it suffices to show that for each $k \in \mathbb{N}$ the set $(A_\alpha \backslash (A_\gamma \cup \bigcup_{\beta \in \Gamma} A_\beta)) \backslash \{0, \dots, k-1\}$ is nonempty. Given $k \in \mathbb{N}$, pick some $p \in \mathcal{G} \cap \mathcal{D}_{\alpha, \Gamma, k}^{(4)}$. Let $i \in (A_\alpha \backslash (X_p \cup \bigcup_{\beta \in \Gamma} A_\beta)) \cap \{k, \dots, \max(X_p)\}$. Clearly $i \in A_\alpha \backslash \bigcup_{\beta \in \Gamma} A_\beta$, $i \notin X_p$ and $k \le i \le \max(X_p)$. If $i \in A_\gamma$, then there is a $q \in \mathcal{G}$ with $i \in X_q$. As $\mathcal{G}$ is a filter, there is a $r \in \mathcal{G}$ with $r \le p, q$. As $r \le q$ it follows that $i \in X_r$. However, then $i \in X_r \backslash X_p$, and so $\min(X_r \backslash X_p) \le i \le \max(X_p)$, contradicting the fact that $r \le p$. Therefore it must be that $i \notin A_\gamma$, and so $i \in (A_\alpha \backslash (A_\gamma \cup \bigcup_{\beta \in \Gamma} A_\beta)) \backslash \{0, \dots, k-1\}$. $\square$

Therefore we may construct a sequence $\{A_\alpha\}_{\alpha < \mathfrak{p}}$ of subsets of $\mathbb{N}$ which satisfies $(1)_\gamma$–$(3)_\gamma$ for each $\gamma < \mathfrak{p}$. It is now easy to show that this is a sequence of infinite subsets of $\mathbb{N}$ such that for each $\{\alpha, \beta\} \in [\mathfrak{p}]^2$ we have

$$\{\alpha, \beta\} \in S \;\Leftrightarrow\; A_\alpha \cap A_\beta \in Fin. \qquad \square$$

Exercise. Show that for every downwards closed set $K \subseteq [\mathfrak{p}]^{<\omega}$ containing all the singletons there is a family A_α $(\alpha < \mathfrak{p})$ of infinite subsets of ω such that for $a \in [\mathfrak{p}]^{<\omega}$,

$$a \in S \text{ iff } \left| \bigcap_{\alpha \in a} A_\alpha \right| = \aleph_0.$$

2.3 Coding sets of ordered pairs

Theorem 12. *Suppose $S \subseteq \mathfrak{p} \times \mathfrak{p}$. Then there are sequences $\{A_\alpha\}_{\alpha < \mathfrak{p}}$, $\{B_\beta\}_{\beta < \mathfrak{p}} \subseteq [\mathbb{N}]^\infty$ such that*

$$(\forall (\alpha, \beta) \in \mathfrak{p} \times \mathfrak{p}) \left[(\alpha, \beta) \in S \iff A_\alpha \cap B_\beta \in Fin \right].$$

Proof. We want to be able to take advantage of the previous theorem, where we coded unordered pairs of ordinals. In order to do that, we need to code the family of ordered pairs of ordinals as a family of unordered pairs of ordinals.

We define an injection from $\mathfrak{p} \times \mathfrak{p}$ into $[\mathfrak{p}]^2$ as follows: First, split $\mathfrak{p}$ into two disjoint sets of the same cardinality, i.e. set $\mathfrak{p} = \Gamma \cup \Delta$, where $|\Gamma| = |\Delta| = \mathfrak{p}$ and $\Gamma \cap \Delta = \emptyset$. Name the relevant bijections

$$\phi : \mathfrak{p} \to \Gamma \text{ and } \psi : \mathfrak{p} \to \Delta.$$

Now define $\Phi : \mathfrak{p} \times \mathfrak{p} \to [\mathfrak{p}]^2$ by

$$\Phi(\alpha, \beta) = \{\phi(\alpha), \psi(\beta)\}.$$

Claim 12.1. *Φ is a well-defined injection.*

Proof of claim. Since $\Gamma \cap \Delta = \emptyset$, it follows that for any $\alpha, \beta \in \mathfrak{p}$, we have $\phi(\alpha) \neq \psi(\beta)$ and so

$$\Phi(\alpha, \beta) = \{\phi(\alpha), \psi(\beta)\} \in [\mathfrak{p}]^2.$$

It also follows that if $\Phi(\alpha, \beta) = \Phi(\alpha', \beta')$ then $(\alpha, \beta) = (\alpha', \beta')$. $\square$

Let $\tilde{S}$ be the result of coding S by Φ. In other words,

$$\tilde{S} = \{\Phi(\alpha, \beta) : (\alpha, \beta) \in S\} \subseteq [\mathfrak{p}]^2.$$

From the previous theorem, we know that there is a sequence $\{C_\xi\}_{\xi < \mathfrak{p}} \subseteq [\mathbb{N}]^\infty$ coding $\tilde{S}$, that is

$$(\forall \{\xi, \eta\} \in [\mathfrak{p}]^2) \left[\{\xi, \eta\} \in \tilde{S} \iff C_\xi \cap C_\eta \in Fin \right].$$

For all $\alpha, \beta < \mathfrak{p}$, define

$$A_\alpha = C_{\phi(\alpha)} \text{ and } B_\beta = C_{\psi(\beta)}.$$

Claim 12.2. $\{A_\alpha\}_{\alpha < \mathfrak{p}}$ and $\{B_\beta\}_{\beta < \mathfrak{p}}$ *are the required sequences that code* S.

Proof of claim. We must show that

$$(\forall (\alpha, \beta) \in \mathfrak{p} \times \mathfrak{p}) \left[(\alpha, \beta) \in S \iff A_\alpha \cap B_\beta \in Fin \right].$$

Suppose $(\alpha, \beta) \in S$. Then we have

$$\{\phi(\alpha), \psi(\beta)\} = \Phi(\alpha, \beta) \in \tilde{S}.$$

It follows that

$$A_\alpha \cap B_\beta = C_{\phi(\alpha)} \cap C_{\psi(\beta)} \in Fin.$$

Conversely, suppose $(\alpha, \beta) \notin S$. Then, since we have already shown that Φ is an injection, we have

$$\{\phi(\alpha), \psi(\beta)\} = \Phi(\alpha, \beta) \notin \tilde{S}.$$

It follows that

$$A_\alpha \cap B_\beta = C_{\phi(\alpha)} \cap C_{\psi(\beta)} \notin Fin.$$ $\square$

$\square$

Corollary. *Suppose $S \subseteq \mathfrak{p} \times \mathfrak{p}$. Then there are sequences*

$$\{C_n\}_{n=0}^{\infty}, \{D_n\}_{n=0}^{\infty} \subseteq \mathcal{P}(\mathfrak{p})$$

such that

$$S = \limsup_{n \to \infty} C_n \times D_n = \bigcap_m \bigcup_{n \geq m} C_n \times D_n.$$

Proof. We need to construct sequences $\{C_n\}_{n=0}^{\infty}$ and $\{D_n\}_{n=0}^{\infty}$ such that an ordered pair is in S iff it is in $C_n \times D_n$ for infinitely many n.

Applying the Theorem, let $\{A_\alpha\}_{\alpha < \mathfrak{p}}$, $\{B_\beta\}_{\beta < \mathfrak{p}} \subseteq [\mathbb{N}]^{\infty}$ be families of infinite subsets of $\mathbb{N}$ coding the *complement* of S. So we have

$$(\forall (\alpha, \beta) \in \mathfrak{p} \times \mathfrak{p}) \, [(\alpha, \beta) \in S \iff A_\alpha \cap B_\beta \notin Fin].$$

For all $n \in \mathbb{N}$, define

$$
\begin{aligned}
C_n &= \{\alpha \in \mathfrak{p} : n \in A_\alpha\}, \\
D_n &= \{\beta \in \mathfrak{p} : n \in B_\beta\}.
\end{aligned}
$$

Now, for all $(\alpha, \beta) \in \mathfrak{p} \times \mathfrak{p}$ we have

$$
\begin{aligned}
(\alpha, \beta) \in S &\iff A_\alpha \cap B_\beta \notin Fin \\
&\iff (\forall m)(\exists n \geq m) \, [n \in A_\alpha \cap B_\beta] \\
&\iff (\forall m)(\exists n \geq m) \, [(\alpha, \beta) \in C_n \times D_n] \\
&\iff (\alpha, \beta) \in \bigcap_m \bigcup_{n \geq m} C_n \times D_n. \qquad \square
\end{aligned}
$$

Corollary. *Suppose $\kappa < \mathfrak{p}$ and $S \subseteq \kappa \times \mathbb{R}$. Then there are sequences $\{C_n\}_{n=0}^{\infty} \subseteq \mathcal{P}(\kappa)$ and $\{D_n\}_{n=0}^{\infty} \subseteq \mathcal{P}(\mathbb{R})$ such that*

$$S = \limsup_{n \to \infty} C_n \times D_n = \bigcap_m \bigcup_{n \geq m} C_n \times D_n.$$

Proof. Once again, we will construct the required sequences by coding the *complement* of S.

First, fix an almost disjoint family $\mathcal{A} = \{A_\alpha\}_{\alpha < \kappa} \subseteq [\mathbb{N}]^{\infty}$. Now, fix $x \in \mathbb{R}$, and let

$$\mathcal{B}_x = \{A_\alpha : \alpha < \kappa, (\alpha, x) \notin S\} \subseteq \mathcal{A}.$$

Since $|\mathcal{A}| = \kappa < \mathfrak{p}$ and $\mathcal{B}_x \subseteq \mathcal{A}$, we can use Theorem 10 to obtain a code for $\mathcal{B}_x$ in $\mathcal{A}$, that is a set $B_x \subseteq \mathbb{N}$ such that

$$(\forall A \in \mathcal{A}) \, [A \in \mathcal{B}_x \iff A \cap B_x \in Fin].$$

We obtain such a code B_x for each $x \in \mathbb{R}$, giving us a collection $\{B_x\}_{x\in\mathbb{R}} \subseteq \mathcal{P}(\mathbb{N})$ such that for all $(\alpha, x) \in \kappa \times \mathbb{R}$ we have

$$
\begin{aligned}
(\alpha, x) \in S &\iff A_\alpha \notin \mathcal{B}_x \\
&\iff A_\alpha \cap B_x \notin Fin.
\end{aligned}
$$

For all $n \in \mathbb{N}$, define

$$
\begin{aligned}
C_n &= \{\alpha < \kappa : n \in A_\alpha\}, \\
D_n &= \{x \in \mathbb{R} : n \in B_x\}.
\end{aligned}
$$

Now, for all $(\alpha, x) \in \kappa \times \mathbb{R}$ we have

$$
\begin{aligned}
(\alpha, x) \in S &\iff A_\alpha \cap B_x \notin Fin \\
&\iff (\forall m)(\exists n \geq m)\,[n \in A_\alpha \cap B_x] \\
&\iff (\forall m)(\exists n \geq m)\,[(\alpha, x) \in C_n \times D_n] \\
&\iff (\alpha, x) \in \bigcap_m \bigcup_{n \geq m} C_n \times D_n. \qquad \square
\end{aligned}
$$

Corollary. *Every subset S of $\omega_1 \times \omega_1$ is in the σ-algebra generated by sets of the form $X \times Y$, for $X, Y \subseteq \omega_1$.*

(In fact, this remains true with ω_1 replaced by any cardinal $\leq \mathfrak{p}$.)

Proof. We know that $\omega_1 \leq \mathfrak{p}$, so the first Corollary to Theorem 12 tells us immediately that S can be written as a countable intersection of countable unions of "rectangles". But of course a σ-algebra is closed under countable unions and intersections, so it follows that S is contained in the σ-algebra generated by the rectangles. $\qquad \square$

The following corollary deals with the nonexistence of a measure on ω_1 satisfying certain properties. The basic definitions relating to measure spaces will be given in Chapter 4. In addition, note that we have the following definition:

Definition. Suppose $(X, \mathcal{B}, \mu)$ is a measure space. We say that μ is *normalized* if $\mu(X) = 1$. In this case, μ is also called a *probability measure*.

For the sake of clarity, note that when we speak of a measure $\mu : \mathcal{P}(X) \to [0, 1]$, we are referring to a measure defined on *all* of $\mathcal{P}(X)$, meaning that all subsets of X are μ-measurable.

Corollary. *There is no normalized countably additive diffuse measure $\mu : \mathcal{P}(\omega_1) \to [0, 1]$.*

Proof. Suppose there were such a measure μ. So all subsets of ω_1 are μ-measurable, and therefore all "rectangles" of the form $X \times Y \subseteq \omega_1 \times \omega_1$ are measurable in the product measure. From the previous corollary, every subset of $\omega_1 \times \omega_1$ is in the σ-algebra generated by the "rectangles", and so every subset of $\omega_1 \times \omega_1$ must be measurable by the product measure $\mu \times \mu$.

In particular, the set

$$S = \{(\alpha, \beta) \in \omega_1 \times \omega_1 : \alpha < \beta\}$$

must be measurable in the product measure. Let $\chi_S : \omega_1 \times \omega_1 \to \{0, 1\}$ be the characteristic function of S, which must be a measurable function. By Fubini's Theorem, the integral

$$\iint_{\omega_1 \times \omega_1} \chi_S \, d\mu \times \mu$$

can be evaluated in two different ways and both must give the same result.

Note that if we fix $\beta \in \omega_1$ then the set $\{\alpha \in \omega_1 : \alpha < \beta\}$ is countable and therefore has μ-measure zero. On the other hand, if we fix $\alpha \in \omega_1$, then the set $\{\beta \in \omega_1 : \alpha < \beta\}$ is the complement of a countable set, and therefore has μ-measure 1, a contradiction. □

2.4 Strong coding

Definition. Let I be any index set of cardinality $\leq \mathfrak{c}$. We say that $S \subseteq I \times I$ admits a strong coding if there exist sequences $\{A_\alpha\}_{\alpha \in I}$ and $\{B_\alpha\}_{\alpha \in I}$ of infinite subsets of $\mathbb{N}$ such that

1. $(\alpha, \beta) \in S \longrightarrow A_\alpha \subseteq^* B_\beta$

2. $(\alpha, \beta) \notin S \longrightarrow A_\alpha \cap B_\beta \in \textit{Fin}.$

Now we will see how the existence of a strong coding for any $S \subseteq I \times I$ is equivalent to the existence of a denumerable base for a family of functions of cardinality $|I|$ (see [20]). Since $\mathfrak{p} \geq \omega_1$, the existence of a base for a family of power $\aleph_1$ follows directly from the following results.

Theorem 13. *Let I any index set of cardinality $\leq \mathfrak{c}$. Then the following are equivalent:*

(1) Every $S \subseteq I \times I$ admits a strong coding.

(2) For every $\{f_\alpha : \alpha \in I\} \subseteq \mathbb{R}^I$ there is $\{g_n : n < \omega\} \subseteq \mathbb{R}^I$ such that every f_α is the pointwise limit of a subsequence of $\{g_n : n < \omega\}$.

Proof. Let us prove (1) implies (2). Let $\mathcal{F} = \{f_\alpha : \alpha \in I\} \subseteq \mathbb{R}^I$. We may assume that $\mathcal{F} \subseteq [0,1]^I$. Now, if $f : I \to [0,1]$, it can be expressed as

$$f = \sum_{i=1}^{\infty} \frac{\phi_f^i(x)}{2^i}$$

where $\phi_f^i(x) = 0, 1$. Let $\mathcal{F}^i = \{\phi_f^i : f \in \mathcal{F}\}$. If $\mathcal{G} = \{g_n^i : n < \omega\} \subseteq \{0,1\}^I$ is a family of functions such that any $\phi \in \mathcal{F}^i$ is the pointwise limit of a subsquence of $\mathcal{G}$, it can be easily checked that any $f \in \mathcal{F}$ is the pointwise limit of functions of the form

$$\sum_{i=1}^{k} \frac{g_{n_i}'(x)}{2^i},$$

where $k \in \mathbb{N}$ and $(n_1, n_2, \ldots, n_k)$ is a finite sequence of natural numbers. Notice $|\mathcal{F}'| \leq |\mathcal{F}|$. Therefore we can also assume that $\mathcal{F} \subseteq \{0,1\}^I$.

Define $S_{\mathcal{F}} \subseteq I \times I$ as follows

$$(\alpha, \beta) \in S_{\mathcal{F}} \text{ iff } f_\alpha(\beta) = 1.$$

Let $\{A_\alpha\}_{\alpha \in I}$ and $\{B_\alpha\}_{\alpha \in I}$ be families of infinite subsets of $\mathbb{N}$ witnessing that $S_{\mathcal{F}}$ admits a strong coding. For $n \in \mathbb{N}$ define $g_n : I \to \{0,1\}$ as follows:

$$g_n(\beta) = 1 \text{ iff } n \in B_\beta.$$

Then we have the following:

Claim. *The sequences $\{A_\alpha\}_{\alpha \in I}$ and $\{B_\alpha\}_{\alpha \in I}$ are a coding for $S_{\mathcal{F}}$ iff for any $\beta \in I$,*

$$\lim_{n \to \infty, n \in A_\alpha} g_n(\beta) = f_\alpha(\beta).$$

Proof. Pick $\beta \in I$ and assume first that $f_\alpha(\beta) = 1$. Then we have that

$$\lim_{n \to \infty, n \in A_\alpha} g_n(\beta) = f_\alpha(\beta) = 1 \text{ iff}$$

$$(\exists N)(\forall n \in A_\alpha) \ (n \geq N \to g_n(\beta) = 1) \text{ iff}$$

$$(\exists N)(\forall n \in A_\alpha) \ (n \geq N \to n \in B_\beta) \text{ iff}$$

$$A_\alpha \subseteq^* B_\beta.$$

Now, if $f_\alpha(\beta) = 0$, we have

$$\lim_{n\to\infty, n\in A_\alpha} g_n(\beta) = f_\alpha(\beta) = 0 \;\; \text{iff}$$

$$(\exists N)(\forall n \in A_\alpha)\;\; (n \geq N \to g_n(\beta) = 0) \;\; \text{iff}$$

$$(\exists N)(\forall n \in A_\alpha)\;\; (n \geq N \to n \notin B_\beta) \;\; \text{iff}$$

$$A_\alpha \cap B_\beta \in Fin.$$

$\square$

To prove (2) implies (1), take $S \subseteq I \times I$ and define $\mathcal{F}_S = \{f_\alpha : \alpha < \mathfrak{m}\}$ as follows:

$$f_\alpha(\beta) = 1 \text{ iff } (\alpha, \beta) \in S.$$

By assuming (2), we can find $\{g_n : n \in \mathbb{N}\} \subseteq 2^I$ such that any f_α is the pointwise limit of a subsequence in $\mathcal{F}_S$. By the preceding claim, we can easily produce a strong coding for S. $\square$

Theorem 14. *Every $S \subseteq \mathfrak{p} \times \mathfrak{p}$ admits a strong coding.*

Proof. We will produce recursively sequences $A_\alpha(\alpha < \mathfrak{p})$ and $B_\alpha(\alpha < \mathfrak{p})$ of infinite subsets of $\mathbb{N}$. Instead of constructing these sequences into the set of natural numbers, we will use as our index set the tree $T = (2^{<\mathbb{N}}, \subseteq)$. Suppose that A_α $(\alpha < \gamma)$ and B_α $(\alpha < \gamma)$ have been defined and they satisfy the following inductive hypothesis

(a) $(\alpha, \beta) \in S \longrightarrow A_\alpha \subseteq^* B_\beta$.

(b) $(\alpha, \beta \notin S) \longrightarrow A_\alpha \cap B_\beta \in Fin$.

(c) A_α is an infinite chain of T (so, $\bigcup A_\alpha$ can be identified with a member of $2^\mathbb{N}$).

(d) $\bigcup A_\alpha \neq \bigcup A_\beta$ for $\alpha \neq \beta < \gamma$.
    For each $\alpha < \gamma$, define $B_\alpha^1 = B_\alpha$ and $B_\alpha^0 = T \setminus B_\alpha$.

(e) For every finite $F \subseteq \gamma$ and $\varepsilon : F \to \{0,1\}$ the set $\bigcap_{\alpha\in F} B_\alpha^{\varepsilon(\alpha)}$ is dense in T.

Let us construct A_γ and B_γ.

Notation. For $x \neq y$ in $2^{\mathbb{N}}$ we will denote $\triangle(x, y) = \min\{n : x(n) \neq y(n)\}$.

To construct B_γ, define $\mathcal{P}_\gamma$ as the collection of all $p = \langle n_p, B_p, F_p \rangle$, where

(f) $B_p : \{0,1\}^{\leq n_p} \to \{0,1\}$.

(g) $F_p \subseteq \gamma$ finite.

(h) $\triangle(\bigcup A_\alpha, \bigcup A_\beta) \leq n_p$ for $\alpha, \beta \in F_p$ such that $(\alpha, \gamma) \in S$ and $(\beta, \gamma) \notin S$.

For $p, q \in \mathcal{P}_\gamma$, set $p \leq q$ iff

(i) $n_p \geq n_q$, $B_q \subseteq B_p$ and $F_q \subseteq F_p$.

The following will provide (a) and (b),

(j) $A_\alpha \restriction_{(n_q, n_p]} \subseteq B_p^1$, for $\alpha \in F_q$ with $(\alpha, \gamma) \in S$.

(k) $A_\alpha \restriction_{(n_q, n_p]} \subseteq B_p^0$, for $\alpha \in F_q$ with $(\alpha, \gamma) \notin S$.

Claim 14.1. $\mathcal{P}_\gamma$ *is a partial order.*

Proof. We will prove that $\mathcal{P}_\gamma$ is transitive. Let $p, q, r \in \mathcal{P}_\gamma$, with $p \leq q \leq r$. We clearly have that $n_p \geq n_r$, $B_p \supseteq B_r$ and $F_p \supseteq F_r$, so, we only have to check (j) and (k) in order to prove $p \leq r$. If $\alpha \in B_r \subseteq B_q$ and $(\alpha, \gamma) \in S$, we have that

$$A_\alpha \restriction_{(n_r, n_q]} \subseteq B_q^1 \subseteq B_p^1$$

and

$$A_\alpha \restriction_{(n_q, n_p]} \subseteq B_p^1,$$

therefore

$$A_\alpha \restriction_{(n_r, n_p]} = A_\alpha \restriction_{(n_r, n_q]} \cup A_\alpha \restriction_{(n_q, n_p]} \subseteq B_p^1.$$

Hence $p \leq r$.

$\square$

Claim 14.2. $\mathcal{P}_\gamma$ *is σ-centered.*

Proof. Let $\{p_\alpha : \alpha < \omega_1\}$ be an uncountable collection of elements of $\mathcal{P}_\gamma$. Since $\{B_p : p \in \mathcal{P}_\gamma\}$ is countable, we must have that there is $p, q \in \mathcal{P}_\gamma$ such that $B_p = B_q$. Hence $r = (n_p, B_p, F_p \cup F_q)$ is a common extension for p and q. Therefore $\mathcal{P}_\gamma$ is σ-centered. $\qquad\square$

For each $n \in \mathbb{N}$ define

$$\mathcal{D}_n = \{p \in \mathcal{P}_\gamma : n_p \geq n\}.$$

If $F \subset \gamma$, $\varepsilon : F \to \{0,1\}$ and $t \in T$, let

$$\mathcal{D}_\varepsilon^t = \{p \in \mathcal{P}_\gamma : (\exists x \in 2^{\leq n_p})(x \supseteq t, B_p(x) = 1),$$
$$(\exists y \in 2^{\leq n_p})(y \supseteq t, B_p(y) = 0),$$
$$\{x, y\} \subseteq \bigcap_{\alpha \in F} B_\alpha^{\varepsilon(\alpha)}\}.$$

For each $\alpha < \gamma$ define

$$\mathcal{D}_\alpha = \{p \in \mathcal{P}_\gamma : \alpha \in F_p\}.$$

Claim 14.3. *Let $n \in \mathbb{N}$, $F \subset \gamma$, $\varepsilon : F \to \{0,1\}$, $t \in T$, and $\alpha \in \gamma$. Then*

(i) $\mathcal{D}_n$ is dense open in $\mathcal{P}_\gamma$.

(ii) $\mathcal{D}_\varepsilon^t$ is dense open in $\mathcal{P}_\gamma$.

(iii) $\mathcal{D}_\alpha = \{p \in \mathcal{P}_\gamma : \alpha \in F_p\}$ is dense open in $\mathcal{P}_\gamma$.

Proof. (i) Let $n \in \mathbb{N}$ and take $q = \langle B_q, P_q, n_q \rangle$ an arbitrary element of $\mathcal{P}_\gamma$. Then (by condition (h)) we have that for $\alpha, \beta \in F_q$ such that $(\alpha, \gamma) \in S$ and $(\alpha, \beta) \notin S$, the following holds

$$\left(\bigcup_{\alpha \in F_q'} A_\alpha\right) \cap \left(\bigcup_{\beta \in F_q''} A_\beta\right) \cap (T \smallsetminus 2^{\leq n_q}) = \emptyset$$

where $F_q' = \{\alpha \in F_q : (\alpha, \gamma) \in S\}$ and $F_q'' = \{\beta \in F_q : (\beta, \gamma) \notin S\}$. Then we can extend q to a condition p as follows: If $n_q \geq n$, we set $p = q$. If $n_q \leq n$ let $p = \langle B_p, F_p, n_p \rangle$, where $F_p = F_q$, $n_p = n$ and $B_p : 2^{<n_p} \to \{0,1\}$ is defined as follows

$$B_p(s) = B_q(s) \text{ if } |s| \in [0, n_q].$$

Now, if $\alpha \in F_q$ and $(\alpha, \gamma) \in S$, we define

$$B_p(s) = 1 \text{ if } |s| \in (n_q, n] \text{ and } s \in A_\alpha.$$

If $\alpha \in F_q$ and $(\alpha, \gamma) \notin S$, we define

$$B_p(s) = 0 \text{ if } |s| \in (n_q, n] \text{ and } s \in A_\alpha$$

we define $B_p(s) = 0$ otherwise. Then it can be proved that $p \in \mathcal{D}_n$ and $p \leq q$.

(ii) Let $F \subset \gamma$ be finite, $\varepsilon : F \to \{0, 1\}$ and $q = \langle B_q, F_q, n_q \rangle$ an arbitrary element of $\mathcal{P}_\gamma$. Let $T^t = \{x \in T : x \supseteq t\}$. By condition (e) we have that $\bigcap_{\alpha \in F} B_\alpha^{\varepsilon(\alpha)}$ is dense in T. Then

$$\left(T^t \cap \bigcap_{\alpha \in F} B_\alpha^{\varepsilon(\alpha)}\right) \setminus \bigcup_{\alpha \in F_q} A_\alpha$$

is also dense in T^t. So, we can find two distinct points x and y in $(T^t \cap \bigcap_{\alpha \in F} B_\alpha^{\varepsilon(\alpha)}) \setminus \bigcup_{\alpha \in F_q} A_\alpha$. Let $n_p = max\{|x|, |y|, n_q\}$ and extend q to a condition p as follows:

Let

$$B_p(s) = B_q(s) \text{ if } |s| \in [0, n_q].$$

Since $x, y \notin \bigcup_{\alpha \in F_q} A_\alpha$, we can define

$$B_p(x) = 1 \text{ and } B_p(y) = 0.$$

If $\alpha \in F_q$ and $(\alpha, \gamma) \in S$, we define

$$B_p(s) = 1 \text{ if } |s| \in (n_q, n_p] \text{ and } s \in A_\alpha.$$

If $\alpha \in F_q$ and $(\alpha, \gamma) \notin S$, we define

$$B_p(s) = 0 \text{ if } |s| \in (n_q, n_p] \text{ and } s \in A_\alpha$$

we define $B_p(s) = 0$ otherwise. Let $F_p = F_q$. Then we have that $p = \langle F_p, B_p, n_p \rangle \in \mathcal{D}_n$ and $p \leq q$.

(iii) Let $\alpha < \gamma$ and $q = \langle B_q, F_q, n_q \rangle$ an arbitrary element of $\mathcal{P}_\gamma$. Then we have that $p = \langle B_q = B_p, F_q \cup \{\alpha\}, n_q = n_p \rangle$ is an extension of q and clearly $p \in \mathcal{D}_\alpha$.

$\square$

Since the number of dense open sets in $\mathcal{P}_\gamma$ defined above is at most

$$\max\{\aleph_0, |\gamma|\} < \mathfrak{m},$$

we can find a generic filter $\mathcal{G} \subseteq \mathcal{P}_\gamma$ intersecting each one of these dense open sets. Define

$$B_\mathcal{G} = \bigcup_{p \in \mathcal{G}} B_p : T \to \{0,1\}$$

and

$$B_\gamma = B_\mathcal{G}^{-1}(1).$$

Claim 14.4. *Let $\alpha < \gamma$. Then the following holds*

(a) *If $(\alpha, \gamma) \in S$, then $A_\alpha \subseteq^* B_\gamma$,*

(b) *If $(\alpha, \gamma) \notin S$, then $A_\alpha \cap A_\gamma \in Fin$.*

Proof.   (a) Suppose that $(\alpha, \gamma) \in S$. Take $p \in \mathcal{G} \cap \mathcal{D}_\alpha$. Then we have that $\alpha \in F_p$. We claim that $A_\alpha \setminus B_\gamma \subseteq B_p^1$. Take $x \in A_\alpha \setminus B_\gamma$. Suppose that $|s| \geq n_p$. Pick $q \in \mathcal{G} \cap \mathcal{D}_n$. It follows that $n_q \geq n$. Since $\mathcal{G}$ is a filter, we can find $r \in \mathcal{G}$ such that $r \leq p, q$. Then by condition (j), we conclude that $x \in B_r(1)$. Hence

$$x \in \bigcup_{p \in \mathcal{G}} B_p^{-1}(1) = B_\gamma,$$

a contradiction. Therefore $|x| \leq n_p$. Since $(\alpha, \gamma) \in S$, by condition (j) again we conclude that $x \in B_p^1$.

(b) Suppose that $(\alpha, \gamma) \notin S$. Take $p \in \mathcal{G} \cap \mathcal{D}_\alpha$. Then we have that $\alpha \in F_p$. Take $x \in A_\alpha \cap B_\gamma$. Since $x \in B_\gamma$, there exists $q \in \mathcal{G}$ such that $x \in B_q^{-1}(1)$. Suppose that $|x| = n \geq n_p$. Find $s \in \mathcal{G} \cap \mathcal{D}_n$, then we have that $n_s \geq n$. Since $\mathcal{G}$ is a filter, we can find $r \in \mathcal{G}$ a common extension of p, q, s. Since $(\alpha, \gamma) \notin S$, we obtain by condition (k) that $B_r(x) = 0$, but this is a contradiction since r extends q. $\qquad\square$

Now we force A_γ. Let $\mathcal{Q}_\gamma$ be the collection of pairs of the form

$$p = (A_p, F_p),$$

where A_p is a finite chain of the tree T and F_p is a finite subset of $\gamma + 1$. Put $p \leq q$ iff

(1) A_p end extends A_q and $F_p \supseteq F_q$.

(m) $A_p \setminus A_q \subseteq B_\alpha$, for every $\alpha \in F_q$ such that $(\gamma, \alpha) \in S$.

(n) $A_p \setminus A_q \subseteq T \setminus B_\alpha$, for every $\alpha \in F_q$ such that $(\gamma, \alpha) \notin S$.

Claim 14.5. $\mathcal{Q}_\gamma$ *is a partial order.*

Proof. We have to prove transitivity. Let $p, q, r \in \mathcal{Q}_\gamma$ such that $p \leq q \leq r$. We clearly have that $A_p \setminus A_r = (A_p \setminus A_q) \cup (A_q \setminus A_r)$ and $F_p \supseteq F_q \supseteq F_r$. Then if $(\gamma, \alpha) \in S$, we have that $A_p \setminus A_r \subseteq B_\alpha$ and if $(\gamma, \alpha) \notin S$, $A_p \setminus A_r \subseteq T \setminus B_\alpha$. Hence $\mathcal{Q}_\gamma$ is transitive. $\square$

Note that if $A_p = A_q$ then $r = (A_p, F_p \cup F_q) \leq p, q$, so $\mathcal{Q}_\gamma$ satisfies the c.c.c.

For each $n \in \omega$ and $\alpha < \gamma$ define

$$\mathcal{D}_n^\alpha = \{p \in \mathcal{Q}_\gamma : (\exists x \in A_p)(|x| > n), (x \not\subseteq \cup A_\alpha)\}.$$

For each $\beta < \gamma + 1$ define

$$\mathcal{E}_\beta = \{p \in \mathcal{Q}_\gamma : \beta \in F_p\}$$

Claim 14.6. *For each $n \in \omega$, $\alpha < \gamma$ and $\beta < \gamma + 1$ the following holds*

 i) $\mathcal{D}_n^\alpha$ *is open dense in* $\mathcal{Q}_\gamma$,

 ii) $\mathcal{E}_\beta$ *is open dense in* $\mathcal{Q}_\gamma$.

Proof. i) Let $n \in \mathbb{N}$, $\alpha < \gamma$ and $q = \langle A_q, F_q \rangle \in \mathcal{Q}_\gamma$. Let $t = \max A_q$. Define $\varepsilon : F_q \to \{0, 1\}$ by $\varepsilon(\alpha) = 1$ iff $(\gamma, \alpha) \in S$. By the inductive hypothesis (e), we have that $\bigcap_{\alpha \in F_q} B_\alpha^{\varepsilon(\alpha)}$ is dense in T, therefore

$$\bigcap_{\alpha \in F_q} B_\alpha^{\varepsilon(\alpha)} \setminus \bigcup A_\alpha$$

is also dense in T. Hence, we can find $x \in \bigcap_{\alpha \in F_q} B_\alpha^{\varepsilon(\alpha)} \setminus \bigcup A_\alpha$ extending t. Let

$$p = (A_q \cup \{x\}, F_q)$$

then p extends q, and clearly p is an element of $\mathcal{D}_n^\alpha$.

ii) Take $\beta < \gamma + 1$ and $q = \langle A_q, F_q \rangle \in \mathcal{Q}_\gamma$ arbitrary. Then we clearly have that $p = \langle A_q, F_q \rangle \cup \{\beta\}$ is an element of $\mathcal{D}_\beta$ and $p \leq q$.

                    $\square$

Since the number of dense subsets defined above is at most $\max\{\aleph_0, |\gamma + 1|\} < \mathfrak{m}$, we can find $\mathcal{G} \subseteq \mathcal{Q}_\gamma$ a generic filter intersecting each one of them. Define

$$A_\gamma = \bigcup_{p \in \mathcal{G}} A_p.$$

For each $n \in \mathbb{N}$, there is $p \in \mathcal{G} \cap \mathcal{D}_n$. Hence A_γ is an infinite chain. Furthermore, if $\alpha < \gamma$ there is $x \in A_p$ such that $x \notin \cup A_\alpha$. Hence $\cup A_\gamma \neq \cup A_\alpha$ for each $\alpha < \gamma$.

Now, take $\alpha < \gamma + 1$. Suppose $(\gamma, \alpha) \in S$. Take $p \in \mathcal{D}_\alpha \cap \mathcal{G}$. Let $x \in A_\gamma \setminus B_\alpha$. Then there is $q \in \mathcal{G}$ such that $x \in A_q$. since $\mathcal{G}$ is a filter, we can find $r \in \mathcal{G}$ which is a common extension for p and q. Then it follows by condition (m) that $A_r \setminus A_p \subset B_\alpha$. Since $x \notin B_\alpha$, we must have that $x \in A_p$. Hence $A_\gamma \setminus B_\alpha \subseteq A_p$, i.e., $A_\gamma \subseteq^* B_\alpha$.

Similarly we prove that if $(\gamma, \alpha) \notin S$, then $A_\gamma \cap A_\alpha$ is finite. Hence A_α $(\alpha < \gamma + 1)$ and B_α $(\alpha < \gamma + 1)$ satisfies the inductive hypothesis (a)-(e). □

Note that the proof shows that every subset of $\mathfrak{p} \times \mathfrak{p}$ admits a strong coding. Note the following consequence which while it mentions no Baire category numbers could have been hardly discovered in any other context.

Corollary 15. *Every subset of $\omega_1 \times \omega_1$ admits a strong coding.*

Proof. This follows directly from the previous result and the fact that $\mathfrak{m} \geq \omega_1$. □

This result was originally proved by Rothberger [20]. However, looking at the proof of this result in [20] one notices that it does not give any larger amount of coding regardless whether $\mathfrak{m}$ or $\mathfrak{p}$ are strictly larger that ω_1.

2.5 Solovay's lemma and its corollaries

The proof of Theorem 10 above gives a bit more. The following result gives us the right generality of that argument.

Theorem 16 (Solovay's Lemma). *Suppose $\mathcal{A}$ and $\mathcal{B}$ are two families of subsets of $\mathbb{N}$, with $|\mathcal{A}|, |\mathcal{B}| < \mathfrak{p}$, such that $A \setminus \bigcup \mathcal{F} \notin Fin$ for every $A \in \mathcal{A}$ and finite $\mathcal{F} \subseteq \mathcal{B}$.*
Then there is an infinite $X \subseteq \mathbb{N}$ such that:

- $X \cap A \notin Fin$ *for every $A \in \mathcal{A}$; and*

- $X \cap B \in Fin$ *for every $B \in \mathcal{B}$.*

Proof. As indicated this follows from an immediate adjustment of the proof
of Theorem 10. To see this, first note that $\mathcal{B}$ here plays the role of $\mathcal{B}$ in that
theorem, while $\mathcal{A} \cup \mathcal{B}$ here plays the role of $\mathcal{A}$ in that theorem. The only
problem is that we have not specified almost disjointness in the hypotheses
of our theorem.

However, it turns out that the almost disjoint condition was used only
once in the proof of that Theorem, namely in the proof of Claim 10.3.
The only consequence of almost disjointness used in that proof was that
$A \setminus \bigcup \mathcal{F}$ is infinite for $A \in \mathcal{A} \setminus \mathcal{B}$ and finite $\mathcal{F} \subseteq \mathcal{B}$, and that is precisely the
hypothesis added to our theorem (with $\mathcal{A}$ playing the role of $\mathcal{A} \setminus \mathcal{B}$).   □

Theorem 17. *Suppose X is a separable metric space, and $\mathcal{K}$ and $\mathcal{F}$ are
two nonempty families of nonempty subsets of X, with $|\mathcal{K}|, |\mathcal{F}| < \mathfrak{p}$, such
that $(\bigcup \mathcal{K}) \cap (\bigcup \mathcal{F}) = \emptyset$. Suppose also that every member of $\mathcal{K}$ is compact,
and every member of $\mathcal{F}$ is closed. Then there is an F_σ-set $Y \subseteq X$ such that
$\bigcup \mathcal{F} \subseteq Y$ and $Y \cap (\bigcup \mathcal{K}) = \emptyset$.*

*Note that this theorem is not symmetric, in the sense that it asserts
the existence of an F_σ-set that includes the closed sets, and a G_δ-set that
includes the compact sets, but not the other way around.*

Proof. For metric spaces, separable is equivalent to second-countable, so
our space X has a countable basis $\mathcal{V}$. We can choose the basis $\mathcal{V}$ to be
closed under finite unions.

Enumerate $\mathcal{V} = \{V_n\}_{n \in \mathbb{N}}$ in such a way that each $V \in \mathcal{V}$ appears as V_n
for infinitely many n.

For each $K \in \mathcal{K}$, let

$$A_K = \{n : K \subseteq V_n\} \in [\mathbb{N}]^\infty.$$

For each $F \in \mathcal{F}$, let

$$B_F = \{n : V_n \cap F \neq \emptyset\} \in [\mathbb{N}]^\infty.$$

Let

$$\mathcal{A} = \{A_K : K \in \mathcal{K}\},$$
$$\mathcal{B} = \{B_F : F \in \mathcal{F}\}.$$

We have $|\mathcal{A}|, |\mathcal{B}| < \mathfrak{p}$.

Claim 17.1. *For all $K \in \mathcal{K}$ and finite $\{F_1, \ldots, F_n\} \subseteq \mathcal{F}$, we have*

$$A_K \setminus \bigcup_{i=1}^n B_{F_i} \notin Fin.$$

Proof of claim. Since each F_i is closed, $\bigcup_{i=1}^{n} F_i$ is a finite union of closed sets and so is closed. Therefore its complement $X \setminus \bigcup_{i=1}^{n} F_i$ is open. Clearly K is included in this complement. So for each $x \in K$ we can find a basis element $V_x \in \mathcal{V}$ such that $x \in V_x \subseteq X \setminus \bigcup_{i=1}^{n} F_i$. We then have

$$K \subseteq \bigcup_{x \in K} V_x \subseteq X \setminus \bigcup_{i=1}^{n} F_i.$$

Since K is compact, the cover must have a finite subcover, and so there exist $x_1, \ldots, x_m \in K$ such that

$$K \subseteq \bigcup_{j=1}^{m} V_{x_j} \subseteq X \setminus \bigcup_{i=1}^{n} F_i.$$

Let $V = \bigcup_{j=1}^{m} V_{x_j}$. Since $\mathcal{V}$ is closed under finite unions, we have $V \in \mathcal{V}$.
Define

$$C = \{l \in \mathbb{N} : V_l = V\} \in [\mathbb{N}]^{\infty}.$$

For each $l \in C$, we have

$$K \subseteq V_l \subseteq X \setminus \bigcup_{i=1}^{n} F_i$$

and therefore $V_l \cap F_i = \emptyset$ for all $i \in \{1, \ldots, n\}$. It follows that $l \in A_K$ and $l \notin B_{F_i}$ for each i, or in other words $l \in A_K \setminus \bigcup_{i=1}^{n} B_{F_i}$. So we have shown that $C \subseteq A_K \setminus \bigcup_{i=1}^{n} B_{F_i}$, and since C is infinite, the claim follows. $\square$

Now we can apply Solovay's Lemma to $\mathcal{A}$ and $\mathcal{B}$, to obtain an infinite $M \subseteq \mathbb{N}$ such that:

- $M \cap A_K \notin Fin$ for every $K \in \mathcal{K}$, and

- $M \cap B_F \in Fin$ for every $F \in \mathcal{F}$.

Define

$$Y = \bigcup_{m=0}^{\infty} \bigcap_{\substack{n \in M \\ n \geq m}} (X \setminus V_n).$$

Claim 17.2. *The set Y satisfies the conclusion of the theorem.*

Proof of claim. Each V_n is a basis element and therefore open, so $X \setminus V_n$ is closed. It follows that

$$\bigcap_{\substack{n \in M \\ n \geq m}} (X \setminus V_n)$$

is an intersection of closed sets and is therefore closed. Then Y is a countable union of closed sets, and is therefore an F_{σ}-set.

To show that $\bigcup \mathcal{F} \subseteq Y$: Suppose $x \in \bigcup \mathcal{F}$. Then $x \in F$ for some $F \in \mathcal{F}$. Since $F \in \mathcal{F}$, we have:

$$M \cap B_F \in Fin$$
$$(\exists m)(\forall n \geq m) \left[n \notin M \cap B_F\right]$$
$$(\exists m)(\forall n \geq m) \left[n \in M \Rightarrow n \notin B_F\right]$$
$$(\exists m) \, (\forall n \geq m, n \in M) \left[n \notin B_F\right]$$
$$(\exists m) \, (\forall n \geq m, n \in M) \left[V_n \cap F = \emptyset\right]$$
$$(\exists m) \, (\forall n \geq m, n \in M) \left[x \in F \Rightarrow x \notin V_n\right].$$

But we know $x \in F$, so we have

$$(\exists m) \, (\forall n \geq m, n \in M) \left[x \notin V_n\right]$$
$$(\exists m) \, (\forall n \geq m, n \in M) \left[x \in X \setminus V_n\right]$$
$$x \in \bigcup_{m=0}^{\infty} \bigcap_{\substack{n \in M \\ n \geq m}} (X \setminus V_n)$$
$$x \in Y.$$

So we have shown that $\bigcup \mathcal{F} \subseteq Y$.

To show that $Y \cap (\bigcup \mathcal{K}) = \emptyset$: Suppose $x \in \bigcup \mathcal{K}$. Then $x \in K$ for some $K \in \mathcal{K}$. Since $K \in \mathcal{K}$, we have:

$$M \cap A_K \notin Fin$$
$$(\forall m)(\exists n \geq m) \left[n \in M \cap A_K\right]$$
$$(\forall m) \, (\exists n \geq m, n \in M) \left[n \in A_K\right]$$
$$(\forall m) \, (\exists n \geq m, n \in M) \left[K \subseteq V_n\right]$$
$$(\forall m) \, (\exists n \geq m, n \in M) \left[x \in K \Rightarrow x \in V_n\right].$$

But we know $x \in K$, so we have

$$(\forall m) \, (\exists n \geq m, n \in M) \left[x \in V_n\right]$$
$$(\forall m) \, (\exists n \geq m, n \in M) \left[x \notin X \setminus V_n\right]$$
$$x \notin \bigcup_{m=0}^{\infty} \bigcap_{\substack{n \in M \\ n \geq m}} (X \setminus V_n)$$
$$x \notin Y.$$

So we have shown that $Y \cap (\bigcup \mathcal{K}) = \emptyset$. $\square$

$\square$

Corollary 18 (Silver's Theorem). *Suppose X is a separable metric space of size $< \mathfrak{p}$. Then every subset of X is F_σ and also G_δ.*

Proof. Given $Z \subseteq X$, define

$$\begin{aligned}
\mathcal{K} &= \{\{z\} : z \notin Z\}, \\
\mathcal{F} &= \{\{z\} : z \in Z\}.
\end{aligned}$$

It is clear that $\mathcal{K}$ and $\mathcal{F}$ satisfy all the hypotheses of the theorem (note that a metric space is Hausdorff and so singletons are closed and compact). The theorem then gives us that Z is F_σ.

Since we have shown that every subset of X is F_σ, it follows that every subset of X is the complement of an F_σ-set, and is therefore G_δ. $\qquad\square$

Corollary 19. *For any $\kappa < \mathfrak{p}$, we have $2^\kappa \leq 2^{\aleph_0}$.*

Proof. If κ is finite, then trivially 2^κ is also finite. So we assume $\aleph_0 \leq \kappa < \mathfrak{p}$.

Let X be any subspace of $\mathbb{R}$ such that $\mathbb{Q} \subseteq X \subseteq \mathbb{R}$ and $|X| = \kappa$. Clearly, $\mathbb{Q}$ is a countable dense subset of X and so X is a separable metric space of size $\kappa < \mathfrak{p}$. The previous corollary tells us that every subset of X is G_δ. Since $|X| = \kappa$, the number of subsets of X is $|\mathcal{P}(X)| = 2^{|X|} = 2^\kappa$.

Now, how many G_δ-sets can X have? For metric spaces, separable is equivalent to second-countable, so our space X has a countable basis $\mathcal{V}$. Each open set in X is a union of a collection of basis elements, and there can be at most $2^{|\mathcal{V}|} = 2^{\aleph_0}$ such collections, so there are at most $2^{\aleph_0}$ open sets in X. Each G_δ-set in X is an intersection of a countable collection of open sets, so there can be at most $\left(2^{\aleph_0}\right)^{\aleph_0} = 2^{\aleph_0}$ such sets.

So we have shown $2^\kappa \leq 2^{\aleph_0}$. $\qquad\square$

Theorem 20. *Suppose X is a separable metric space, and $\mathcal{F}$ is a family of closed nowhere-dense subsets of X, with $|\mathcal{F}| < \mathfrak{p}$. Then there exists a sequence $\{N_k\}_{k=0}^\infty$ of (closed) nowhere-dense subsets of X such that for every $F \in \mathcal{F}$ there is some $k \in \mathbb{N}$ such that $F \subseteq N_k$.*

Proof. For metric spaces, separable is equivalent to second-countable, so our space X has a countable basis $\mathcal{V}$.

Enumerate $\mathcal{V} = \{V_n\}_{n \in \mathbb{N}}$ in such a way that each $V \in \mathcal{V}$ appears as V_n for infinitely many n.

For each $k \in \mathbb{N}$, let

$$A_k = \{n : V_n \subseteq V_k\} \in [\mathbb{N}]^\infty.$$

For each $F \in \mathcal{F}$, let

$$B_F = \{n : V_n \cap F \neq \emptyset\} \in [\mathbb{N}]^\infty.$$

Let

$$\begin{aligned} \mathcal{A} &= \{A_k : k \in \mathbb{N}\}, \\ \mathcal{B} &= \{B_F : F \in \mathcal{F}\}. \end{aligned}$$

We have $|\mathcal{A}|, |\mathcal{B}| < \mathfrak{p}$.

Claim 20.1. *For all $k \in \mathbb{N}$ and finite $\{F_1, \ldots, F_j\} \subseteq \mathcal{F}$, we have*

$$A_k \setminus \bigcup_{i=1}^{j} B_{F_i} \notin Fin.$$

Proof of claim. Since each F_i is closed, $\bigcup_{i=1}^{j} F_i$ is a finite union of closed sets and so is closed. Since each F_i is nowhere-dense, $\bigcup_{i=1}^{j} F_i$ is a finite union of nowhere-dense sets and so is nowhere-dense. In particular, $\bigcup_{i=1}^{j} F_i$ cannot include the basis element V_k. So $V_k \setminus \bigcup_{i=1}^{j} F_i$ is a non-empty open set and therefore must include a basis element $V \in \mathcal{V}$.
  Define

$$C = \{l \in \mathbb{N} : V_l = V\} \in [\mathbb{N}]^{\infty}.$$

For each $l \in C$, we have

$$V_l \subseteq V_k \setminus \bigcup_{i=1}^{j} F_i$$

and therefore $V_l \subseteq V_k$ and $V_l \cap F_i = \emptyset$ for all $i \in \{1, \ldots, j\}$. It follows that $l \in A_K$ and $l \notin B_{F_i}$ for each i, or in other words $l \in A_K \setminus \bigcup_{i=1}^{j} B_{F_i}$. So we have shown that $C \subseteq A_K \setminus \bigcup_{i=1}^{j} B_{F_i}$, and since C is infinite, the claim follows. $\square$

Now we can apply Solovay's Lemma to $\mathcal{A}$ and $\mathcal{B}$, to obtain an infinite $M \subseteq \mathbb{N}$ such that:

- $M \cap A_k \notin Fin$ for every $k \in \mathbb{N}$, and

- $M \cap B_F \in Fin$ for every $F \in \mathcal{F}$.

For each $m \in \mathbb{N}$, let

$$N_m = X \setminus \bigcup_{\substack{n \in M \\ n \geq m}} V_n.$$

Claim 20.2. *The sequence $\{N_m\}_{m=0}^{\infty}$ satisfies the conclusion of the theorem.*

Proof of claim. First, notice that each N_m is closed, since it is the complement of a union of basis elements.

To show that each N_m is nowhere-dense: Fix $m \in \mathbb{N}$. We know that N_m is closed, so to show that it is nowhere-dense, it suffices to show that N_m cannot include any basis element. Fix $k \in \mathbb{N}$ and we will show that N_m cannot include the basis element V_k. We have:

$$M \cap A_k \notin Fin$$

and so for our m which is already fixed, we have

$$(\exists n \geq m)\,[n \in M \cap A_k]$$

$$(\exists n \geq m, n \in M)\,[n \in A_k]$$

$$(\exists n \geq m, n \in M)\,[V_n \subseteq V_k]\,.$$

Since we know V_n is a basis element and therefore nonempty, we then have

$$(\exists n \geq m, n \in M)\,[V_n \cap V_k \neq \emptyset]$$

$$\left(\bigcup_{\substack{n \in M \\ n \geq m}} V_n \right) \cap V_k \neq \emptyset$$

$$V_k \not\subseteq \left(X \setminus \bigcup_{\substack{n \in M \\ n \geq m}} V_n \right) = N_m.$$

So we have shown that N_m is nowhere-dense.

To show that each F is included in some N_m: Fix $F \in \mathcal{F}$. Then we have:

$$M \cap B_F \in Fin$$

$$(\exists m)(\forall n \geq m)\,[n \notin M \cap B_F]\,.$$

Fixing such an m, we will show that $F \subseteq N_m$:

$$(\forall n \geq m)\,[n \notin M \cap B_F]$$
$$(\forall n \geq m)\,[n \in M \Rightarrow n \notin B_F]$$
$$(\forall n \geq m, n \in M)\,[n \notin B_F]$$
$$(\forall n \geq m, n \in M)\,[V_n \cap F = \emptyset]$$
$$(\forall n \geq m, n \in M)\,[x \in F \Rightarrow x \notin V_n]$$
$$x \in F \Rightarrow (\forall n \geq m, n \in M)\,[x \notin V_n]$$
$$x \in F \Rightarrow x \notin \bigcup_{\substack{n \in M \\ n \geq m}} V_n$$
$$x \in F \Rightarrow x \in \left(X \setminus \bigcup_{\substack{n \in M \\ n \geq m}} V_n \right) = N_m$$
$$F \subseteq N_m. \qquad \square$$

$$\square$$

Corollary. *The intersection of fewer than $\mathfrak{p}$ dense-open subsets of a complete separable metric space includes a dense G_δ-set.*

Proof. Suppose X is a complete separable metric space, and suppose $\mathcal{D}$ is a family of fewer than $\mathfrak{p}$ dense-open subsets of X. Let

$$\mathcal{F} = \{X \setminus D : D \in \mathcal{D}\}.$$

So $\mathcal{F}$ is a family of fewer than $\mathfrak{p}$ closed nowhere-dense subsets of X.

From the theorem, it follows that there exists a sequence $\{N_k\}_{k=0}^\infty$ of closed nowhere-dense subsets of X such that for every $F \in \mathcal{F}$ there is some $k \in \mathbb{N}$ such that $F \subseteq N_k$. This means that we have

$$\bigcup \mathcal{F} \subseteq \bigcup_{k \in \mathbb{N}} N_k.$$

Taking complements and reversing the direction of inclusion, we have

$$X \setminus \bigcup \mathcal{F} \supseteq X \setminus \bigcup_{k \in \mathbb{N}} N_k$$
$$\bigcap_{F \in \mathcal{F}} (X \setminus F) \supseteq \bigcap_{k \in \mathbb{N}} (X \setminus N_k)$$
$$\bigcap \mathcal{D} \supseteq \bigcap_{k \in \mathbb{N}} (X \setminus N_k).$$

Let $G = \bigcap_{k \in \mathbb{N}} (X \setminus N_k)$. So we have

$$\bigcap \mathcal{D} \supseteq G.$$

Now, each N_k is closed and nowhere-dense, so its complement $X \setminus N_k$ is dense-open. G is a countable intersection of open sets and is therefore a G_δ-set. Also, since X is a complete metric space, the Baire Category Theorem applies, and so since G is a countable intersection of dense-open sets, it must be dense. So we have found our dense G_δ-set $G \subseteq \bigcap \mathcal{D}$. $\qquad \square$

Chapter 3

Consequences in Descriptive Set Theory

In this Chapter we examine the influence of the Baire category assumption $\mathfrak{p} > \omega_1$ on the structure of co-analytic set of reals.

3.1 Borel isomorphisms between Polish spaces

We will use the following standard facts about Polish spaces (see [12] and [15]).

Recall that a Polish space is a separable topological space whose topology can be defined by a complete metric.

Theorem 21. *Let X and Y be Polish spaces, $A \subseteq X$ a Borel set and $f : A \to Y$ an injective function. Then*

1. *If f is continuous, then $f''A$ is Borel.*

2. *f is Borel measurable iff its graph is a Borel subset of $X \times Y$.*

3. *If f is Borel measurable, it is a Borel isomorphism between A and $f''A$.*

4. *If X is Polish and $A \subseteq X$ is an uncountable Borel set, then it contains an homeomorphic copy of $2^{\mathbb{N}}$. In particular $|A| = \mathfrak{c}$. If X and Y are Polish spaces and $A \subset X$, $B \subseteq Y$ are uncountable Borel sets, there is a Borel isomorphism $f : A \to B$.*

3.2 Analytic and co-analytic sets

Definition. Let X be a Polish space. A subset A of X is analytic if there exists a Polish space Y and a continuous function $f : Y \to X$ such that $f''Y = A$.

It can be proved (see for example [12]) that given a Polish space X, there exists $F \subseteq \omega^\omega$ a closed subset and a continuous bijection $h : F \to X$. Hence in the previous definition we can substitute Y by the space ω^ω.

Definition. Let X be a Polish space and $A \subseteq X$. We say that A is co-analytic if $X \smallsetminus A$ is analytic.

It can be shown that analytic sets in a Polish space X have some "regularity" properties, for example the perfect set property and the Baire property. Furthermore we have the following

Theorem 22. *If X is a Polish space, there is an analytic set $A \subset X \times \mathbb{N}^{\mathbb{N}}$ such that the analytic subsets of X are precisely the horizontal sections of A.*

Theorem 23. *Let X and Y be Polish spaces. Then*

1. *The intersection and the union of any sequence of co-analytic sets in X is co-analytic.*

2. *If $f : X \to Y$ is a Borel function and $F \subseteq Y$ is co-analytic, then $f^{-1}[F]$ is co-analytic in X.*

Theorem 24. *Any analytic or co-analytic set in a Polish space is expressible as the union of ω_1 Borel sets.*

Theorem 25. *Suppose X and Y are uncountable separable complete metric spaces. Let $A \subseteq X$ and $B \subseteq Y$ be subspaces of cardinality $< \mathfrak{p}$ and let $f : A \to B$ be any bijection. Then there is a Borel homeomorphism $\tilde{f} : X \to Y$ such that $\tilde{f} \restriction_A = f$.*

Proof. By Theorem 21(4), we have in particular that any two uncountable Polish spaces are Borel isomorphic. Then we can assume that $X = Y = 2^{\mathbb{N}}$. By Silver's theorem (Corollary 18), every subset of A is relatively F_σ in A and the same holds for B. So, for any $n \in \mathbb{N}$, there exist A_n and B_n F_σ-subsets of $2^{\mathbb{N}}$ such that

1. $A_n \cap A = \{x \in A : f(x)(n) = 1\}$ and

2. $B_n \cap B = \{x \in B : f^{-1}(x)(n) = 1\}$.

Define $g_i : 2^{\mathbb{N}} \to 2^{\mathbb{N}}$ for $i = 1, 2$ as follows: $g_0(x)(n) = 1$ iff $x \in A_n$ and $g_1(y)(n) = 1$ iff $y \in B_n$.

Notice that the inverse image under both g_0 and g_1 of a basic open set in $2^{\mathbb{N}}$ is a Boolean combination of F_σ, therefore they are Borel subsets of $2^{\mathbb{N}}$. Hence g_0 and g_1 are Borel maps and clearly $g_0 \upharpoonright_A = f$ and $g_1 \upharpoonright_B = f^{-1}$.

Let

$$G = \{(x, y) \in 2^{\mathbb{N}} \times 2^{\mathbb{N}} : g_0(x) = y \ g_1^{-1}(y) = x\}.$$

Then by Theorem 21(2), G is a Borel subset of $2^\omega \times 2^\omega$. Notice that the proyections π_0 and π_1 are one to one on G. Therefore $\widetilde{A} = \pi_0'' G$ and $\widetilde{B} = \pi_1'' G$ are Borel subsets of $2^{\mathbb{N}}$ (Theorem 21(1)). Furthermore $A \subseteq \widetilde{A}$ and $B \subseteq \widetilde{B}$. Also, we can assume that $2^{\mathbb{N}} \setminus \widetilde{A}$ and $2^{\mathbb{N}} \setminus \widetilde{B}$ are both uncountable. Otherwise we can shrink them so that this happens.

By Theorem 21(4), we can find a Borel homeomorphism $h : 2^{\mathbb{N}} \setminus \widetilde{A} \to 2^{\mathbb{N}} \setminus \widetilde{B}$. Since $g = g_0 \upharpoonright_{\widetilde{A}} : \widetilde{A} \to \widetilde{B}$ is Borel homeomorphism, we have that $\widetilde{f} = h \cup g_0$ is a Borel homeomorphism extending f. $\qquad\square$

3.3   Analytic and co-analytic sets under $\mathfrak{p} > \omega_1$

Corollary (Martin and Solovay). *If* $\mathfrak{p} > \omega_1$, *then the following are equivalent*

1. *There is an uncountable co-analytic set without a perfect subset.*

2. *Every set of size at most $\aleph_1$ in a complete separable metric space is co-analytic.*

3. *Every union of at most $\aleph_1$ Borel set is a continuous image of a co-analytic set.*

Proof. (2) implies (1) is trivial, since every set of size $\aleph_1$ contains no perfect set (we are assuming $\mathfrak{p} > \omega_1$). Now, we prove (1) implies (2). Let C be a co-analytic set without a perfect subset in a Polish space X. Then by Theorem 24, C is the union of $\aleph_1$ many Borel sets, i.e.,

$$C = \bigcup_{\alpha < \omega_1} B_\alpha$$

where each B_α is Borel. We must have that $|B_\alpha| \leq \omega_1$, otherwise B_α would include a perfect subset. Hence $|C| = \aleph_1$. By Theorem 25, if B is any other subset of size $\aleph_1$ in a separable metric space Y, there is a Borel homeomorphism $\widetilde{f} : X \to Y$ such that $\widetilde{f}''C = B$. Since Borel 1-1 images of co-analytic sets is co-analytic, we conclude that B is co-analytic.

(3) implies (2) is trivial. Let us prove (2) implies (3).

Let $B = \bigcup_{\alpha < \omega} B_\alpha \subseteq X$ where each B_α is Borel and X is a Polish space. Since Borel sets are analytic, by Theorem 22, there is an analytic set $A \subseteq X$ such that for every Borel set $B \subset X$ there is $y \in \mathbb{N}^\mathbb{N}$ such that

$$A^y = \{x \in X : (x, y) \in A\} = B.$$

For each α pick $y_\alpha \in \mathbb{N}^\mathbb{N}$ such that $B_\alpha = A^{y_\alpha}$. By (2), and Theorem 23(1), we have that the set $\{y_\alpha : \alpha < \omega_1\}$ is co-analytic and $\bigcup_{\alpha < \omega_1} B_\alpha = \pi_0''(\pi_1^{-1}(Y) \cap A)$ is co-analytic, since $\pi_1^{-1}(Y) \cap A$ is co-analytic. $\qquad\square$

Chapter 4

Consequences in Measure Theory

4.1 Measure spaces

Definition. $(X,\mathcal{B},\mu)$ is a *measure space* if X is a topological space, $\mathcal{B}$ a σ-field of subsets of X including all open sets, μ is a σ-additive function taking $\mathcal{B} \to [0,\infty]$. Such μ is said to be a *measure* on X.

Definition. 1. A measure μ is *diffuse* if for any $x \in X$, $\mu(\overline{\{x\}})=0$.

2. μ is *locally finite* if for any $x \in X$, there exists a set U open in X such that $x \in U$ and $\mu(U) < \infty$.

3. μ is a *Radon measure* if for any $B \in \mathcal{B}$, $\mu(B) = sup\{\mu(K) : K \subset B, K \in \mathcal{B}, K \text{ is compact}\}$.

4. μ is *regular* (or *outer regular*) if for any $B \in \mathcal{B}$, $\mu(B) = inf\{\mu(U) : B \subseteq U \text{ and } U \text{ is open}\}$.

5. μ is *finite* if $\mu(X)$ is finite.

6. μ is *σ-finite* if $X = \bigcup_{n \in \omega} X_n$ such that for any n, $\mu(X_n)$ is finite.

We have the following theorem.

Theorem 26. *Let $(X,\mathcal{B},\mu)$ be a locally finite Radon measure space. Let $\mathcal{N}$ be a collection of size $< \mathfrak{m}$ of measure zero subsets of X. Then every compact subset of $\bigcup \mathcal{N}$ which belongs to $\mathcal{B}$ has measure zero.*

Proof. Suppose not and let $K \in \mathcal{B}$ be a compact set such that K has positive measure and $K \subseteq \bigcup \mathcal{N}$. By local finiteness of X and compactness of K, we

45

know $\mu(K)$ is finite. Now to get a contradiction we want to force a point in K so that it does not belong to any member of $\mathcal{N}$.

Let $\mathcal{P}$ be the collection $\{H \subseteq K : H \text{ is compact}, H \in \mathcal{B}, \mu(H) > 0\}$ ordered by $\subseteq$.

Claim 26.1. $\mathcal{P}$ *is ccc.*

Proof of claim. Let $\mathcal{H} \subseteq \mathcal{P}$ be an uncountable collection, we would like to find $H_0 \neq H_1$ (both in $\mathcal{H}$ such that $\mu(H_0 \cap H_1) > 0$. Suppose we cannot find such H_0 and H_1, i.e., for any distinct $H_0, H_1 \in \mathcal{H}$, $\mu(H_0 \cap H_1) = 0$. Since $\mathbb{Q}$ is dense and countable in $\mathbb{R}$ and $\mathcal{H}$ is uncountable, there is $\epsilon > 0$ and an uncountable $\mathcal{K} \subseteq \mathcal{H}$ such that $\mu(H) > \epsilon$ for any H$\in \mathcal{K}$.

By finiteness of $\mu(K)$, let $n = \lceil \mu(K)/\epsilon \rceil$, and $H_0, ..., H_n$ be $n+1$ distinct elements in $\mathcal{K}$.

Note that, since $\mu(H_i \cap H_j) = 0$ for all $i \neq j$,

$$\mu(\bigcup_{i=0}^{n} H_i)$$

$$= \mu(\bigcup_{i=0}^{n}(H_i \setminus \bigcup_{j<i} H_j))$$

$$= \sum_{i=0}^{n} \mu(H_i \setminus \bigcup_{j<i} H_j)$$

$$= \sum_{i=0}^{n}(\mu(H_i) - \sum_{j<i} \mu(H_i \cap H_j))$$

$$= \sum_{i=0}^{n} \mu(H_i)$$

$$> (n+1)\epsilon$$

$$> \mu(K)$$

which is a contradiction. Therefore we must be able to find distinct $H_0, H_1 \in \mathcal{H}$ such that $\mu(H_0 \cap H_1) > 0$. $\square$

To continue our proof we now construct our dense sets. For $N \in \mathcal{N}$ we define

$$\mathcal{D}_N = \{H \in \mathcal{P} : H \cap N = \emptyset\}.$$

Claim 26.2. $\mathcal{D}_N$ *is dense in* $\mathcal{P}$ *for all* $N \in \mathcal{N}$.

Proof. Let $L \in \mathcal{P}$ be given. Since $\mu(N)=0$, $\mu(L \setminus N) > 0$. By the Radon property there is a compact $H \in \mathcal{B}$ such that $H \subseteq L \setminus N$ and $\mu(H) > 0$, but then $H \in \mathcal{D}_N$ and $H \leq L$. $\square$

Now since $|\mathcal{N}| < \mathfrak{m}$, let $\mathcal{G} \subseteq \mathcal{P}$ be a filter intersecting all dense sets of the form $\mathcal{D}_N (N \in \mathcal{N})$.

By compactness, $\bigcap \mathcal{G} \neq \emptyset$, a contradiction since $\bigcap \mathcal{G} \subseteq K \backslash \bigcup \mathcal{N} = \emptyset$. $\square$

In order for $\bigcup \mathcal{N}$ to have measure zero we need a stronger hypothesis:

Theorem 27. *Let $(X, \mathcal{B}, \mu)$ be a σ-finite outer-regular Radon Hausdorff space with separable $L_1(X, \mathcal{B}, \mu)$ (i.e., $(\mathcal{B}/\mathcal{N}_\mu, d_\mu)$ is a separable metric space where $\mathcal{N}_\mu$ is the collection of measure zero sets and $d_\mu(A, B) = \mu(A \Delta B)$). If $\mathcal{N}$ is a collection of size $< \mathfrak{m}$ of measure zero subsets of X, then $\bigcup \mathcal{N}$ has measure zero.*

Proof. Since $X = \bigcup_{n \in \mathbb{N}} X_n$ where each X_n has finite measure, and μ is σ-additive, we may assume $\mu(X) < \infty$. Therefore it suffices to show that for any $\epsilon < \mu(X)$ there is a compact $K \in \mathcal{B}$ such that $\mu(K) > \epsilon$ and $K \cap (\bigcup \mathcal{N}) = \emptyset$.

We define a partial order $\mathcal{P}_\epsilon$ (where $\epsilon > 0$) as follows:

$$\mathcal{P}_\epsilon = \{K \in \mathcal{B} : K \text{ compact and } \mu(K) > \epsilon\}, \text{ ordered by inclusion.}$$

Claim 27.1. *$\mathcal{P}_\epsilon$ satisfies the ccc.*

Proof. Let $\mathcal{K} \subseteq \mathcal{P}$ be an uncountable family. Since $\mathbb{Q}$ is dense in $\mathbb{R}$ there is a $\delta > 0$ and an uncountable subset $\mathcal{K}_0 \subseteq \mathcal{K}$ such that $\mu(K) > \epsilon + \delta$ for any $K \in \mathcal{K}_0$. By our assumption that $L_1(X, \mathcal{B}, \mu)$ is separable, fix a sequence $\{A_n\}_{n=0}^\infty \subseteq \mathcal{B}$ of sets such that for any $B \in \mathcal{B}$ and $\lambda > 0$, there exists n such that $\mu(B \Delta A_n) < \lambda$.

For $n \in \mathbb{N}$ we define

$$\mathcal{H}_n = \{K \in \mathcal{K}_0 : \mu(A_n \Delta K) < \delta/3\}.$$

Therefore $\mathcal{K}_0 = \bigcup_{n \in \mathbb{N}} \mathcal{H}_n$. As a result there is an n such that $\mathcal{H}_n$ is uncountable. Fix such an n and let H, K be distinct elements of $\mathcal{H}_n$.

We would like to show $\mu(H \cap K) > \epsilon$. Note that since

$$H \Delta K \subseteq (H \Delta A_n) \cup (K \Delta A_n),$$

we have

$$\mu(H \Delta K) \leq \mu(H \Delta A_n) + \mu(K \Delta A_n) \leq 2\delta/3.$$

Also, since $\mu(H \cup K) = \mu(H \cap K) + \mu(H \Delta K)$, we have

$$\mu(H \cap K) = \mu(H \cup K) - \mu(H \Delta K) \geq \epsilon + \delta - 2\delta/3 > \epsilon.$$

Therefore $H \cap K$ belongs to $\mathcal{P}_\epsilon$ and is $\leq$ both H and K. $\square$

Now for $N \in \mathcal{N}$, we let

$$\mathcal{D}_N = \{K \in \mathcal{P}_\epsilon : K \cap N = \emptyset\}.$$

Claim 27.2. *$\mathcal{D}_N$ is dense in $\mathcal{P}_\epsilon$ for any $N \in \mathcal{N}$.*

Proof. Fix $H \in \mathcal{P}_\epsilon$ and $N \in \mathcal{N}$. Then $\mu(H\backslash N) > \epsilon$. By the Radon property we have a compact $K \in \mathcal{B}$ such that $K \subseteq H\backslash N$ and $\mu(K) > \epsilon$. This implies $K \in \mathcal{D}_N$ and $K \leq H$.   □

Let $\mathcal{G} \subseteq \mathcal{P}_\epsilon$ be a filter such that $\mathcal{G} \cap \mathcal{D}_N \neq \emptyset$ for all $N \in \mathcal{N}$, and $H = \bigcap \mathcal{G}$. Then $H \cap (\bigcup \mathcal{N}) = \emptyset$ and $\mu(H) \geq \epsilon$, as required.   □

4.2   More on measure spaces

Using the two lemmas from the previous section we prove the following theorem.

Theorem 28. *If* $\mathfrak{m} > \omega_1$, *every locally finite outer-regular Radon measure space* $(X, \mathcal{B}, \mu)$ *is* σ-*finite.*

Proof. First we observe that every compact $K \in \mathcal{B}$ of positive measure includes a compact subset K_0 such that $\mu(K_0) = \mu(K)$ and K_0 is μ-*supporting* (i.e., for any open set U such that $U \cap K_0 \neq \emptyset$, $\mu(U \cap K_0) > 0$). In fact we only need to let

$$K_0 = K\backslash \bigcup \{U : U \text{ open and } \mu(U \cap K) = 0\}.$$

K_0 is compact since it is a closed subset of K, and $\mu(K_0) = \mu(K)$ follows from the fact that our measure is Radon.

Now we let $\mathcal{K}$ be a maximal family of pairwise-disjoint μ-supporting compact sets. Note that $\mu(X\backslash \bigcup \mathcal{K}) = 0$ by the fact that our measure is Radon. Therefore if $\mathcal{K}$ is countable, $(X, \mathcal{B}, \mu)$ is σ-finite. We now assume $\mathcal{K}$ is uncountable and show that μ cannot be both Radon and regular, leading to a contradiction.

Note that for any $K \in \mathcal{K}$, by compactness and local finiteness we may fix an open set $U(K) \supseteq K$ such that $\mu(U(K)) < \infty$. We let $\mathcal{P}$ be the collection of all finite subsets $p \subset \mathcal{K}$ such that for any $K \in p$,

$$F_p(K) = K\backslash \bigcup \{U(H) : H \in p, H \neq K\}$$

has positive measure. $\mathcal{P}$ is ordered by reverse inclusion. Before we move on, we define *Knaster's property*, which is stronger than the ccc.

Definition. A poset $\mathcal{B}$ has the Knaster's property (property K) if for every uncountable sequence b_ξ ($\xi < \omega_1$) in $\mathcal{B}$ there is an uncountable subsequence b_ξ ($\xi \in \Gamma \subseteq \omega_1$) such that $\neg b_\xi \perp b_\eta$ for all $\xi, \eta \in \Gamma$.

Claim 28.1. $\mathcal{P}$ *has Knaster's property.*

Before we prove the claim, note that by the Δ-system lemma, we may assume:

1. There is p such that for any $p_1, p_2 \in \mathcal{P}$, $p_1 \cap p_2 = p$. In fact, by removing p from each member of $\mathcal{P}$, we may assume that for any $p_1, p_2 \in \mathcal{P}$, $p_1 \cap p_2 = \emptyset$.

2. There is $n \in \mathbb{N}$ such that for all $q \in \mathbb{P}$, $|q \backslash p| = n$. (We restrict $\mathcal{P}$ to those elements which satisfy the equation above. We can choose an n so that the refinement will have the same cardinality as $\mathcal{P}$ since $\mathbb{N}$ is countable.)

3. There is $\epsilon \in \mathbb{R}$ such that $\epsilon > 0$ and $\mu(F_{p_\alpha}(K)) > \epsilon$ for all $K \in p_\alpha$ and $p_\alpha \in \mathcal{P}$.

4. There is $M \in \mathbb{R}$ such that $\mu(\bigcup\{U(K) : K \in p_\alpha\}) < M$ for all $p_\alpha \in \mathcal{P}$.

We can make Assumptions 3 and 4 since $\mathbb{Q}$ is countable and dense in $\mathbb{R}$.

To prove Claim 28.1 first we prove the following:

Lemma. *Suppose $\mathcal{P} = p_\alpha$ ($\alpha \in \omega_1$). We may refine $\mathcal{P}$ to $\mathcal{P}' = p_{\alpha_\xi}$ ($\xi \in \omega_1$) such that for any $\xi, \eta \in \omega_1$, where $\xi < \eta$,*

$$(\bigcup\{U(K) : K \in p_{\alpha_\xi}\}) \cap (\bigcup p_{\alpha_\eta}) = \emptyset.$$

Note that by (1) above we may assume the root of $\mathcal{P}$ is empty.

Proof. We construct $\mathcal{P}'$ by induction. First we let $p_{\alpha_0} = p_0$. Now suppose p_{α_ξ} ($\xi < \eta$) has been constructed, we construct p_{α_η}.

Note that if for any $\beta \geq \eta$ we have

$$\bigcup\{U(K) : K \in p_{\alpha_\xi}, \xi \in \eta\} \cap (\bigcup p_\beta) \neq \emptyset,$$

then there exists $\xi \in \omega_1$ such that

$$S_\xi := \{\beta < \omega_1 : \eta \leq \beta, \bigcup\{U(K) : K \in p_{\alpha_\xi}\} \cap (\bigcup p_\beta) \neq \emptyset\}$$

is uncountable. Note that for any $\beta \in S_\xi$, there is $H_\beta \in p_\beta$ such that

$$\mu(H_\beta \cap \bigcup\{U(K) : K \in p_{\alpha_\xi}\}) > 0$$

and $\beta_1 \neq \beta_2 \Rightarrow H_{\beta_1} \neq H_{\beta_2}$ since the root of $\mathcal{P}$ is empty. Since S_ξ is uncountable, we can refine S_ξ to a set of the same cardinality such that there is an ϵ such that for any β in the refinement, we have

$$\mu(H_\beta \cap \bigcup\{U(K) : K \in p_{\alpha_\xi}\}) > \epsilon,$$

contradicting the fact that $\bigcup\{U(K) : K \in p_{\alpha_\xi}\}$ has finite measure. Therefore there exits $\beta \geq \eta$ such that

$$\bigcup\{U(K) : K \in p_{\alpha_\xi}, \xi < \eta\} \cap (\bigcup p_\beta) = \emptyset.$$

We let α_η be such a β. $\square$

We prove another claim before we prove Claim 28.1. We let

$$I(\beta) = \{\alpha < \beta : p_\alpha \cup p_\beta \notin \mathcal{P}\}.$$

Claim 28.2. $|I(\beta)| \le m$ where m is a natural number such that $m\epsilon > M$. (M and ϵ are as given in assumptions (3) and (4) above.)

Proof. To see this we note that $\alpha \in I(\beta)$ iff there exists $K \in p_\alpha \cup p_\beta$ such that $\mu(F_{p_\alpha}(K) \cap F_{p_\beta}(K)) = 0$.

By our previous lemma we must have $K \in p_\alpha$. Therefore there exists $K_\alpha \in p_\alpha$ such that

$$F_{p_\alpha}(K_\alpha) \subseteq_{a.e.} \bigcup\{U(H) : H \in p_\beta\}$$

(where $A \subseteq_{a.e.} B$ means $A\backslash B$ has measure zero). Therefore

$$|I(\beta)| \cdot \epsilon = \mu(\bigcup\{F_{p_\alpha}(K_\alpha) : \alpha \in I(\beta)\}) \le \mu(\bigcup\{U(H) : H \in p_\beta\}) < M.$$

$\square$

To prove Claim 28.1, we observe that by the free-set lemma there is an uncountable $\Gamma \subseteq \omega_1$ such that for any distinct $\xi, \eta \in \Gamma$, $\xi \notin I(\eta)$. Therefore $\{p_\alpha\}(\alpha \in \Gamma)$ are pairwise compatible in $\mathcal{P}$, as required.

To continue our proof that μ cannot be both Radon and regular, we recall the following lemma from a previous class:

Lemma. *If* $\mathfrak{m} > \omega_1$, *every uncountable ccc poset* $\mathcal{P}$ *contains an uncountable centred subset* $\mathcal{G}$.

$\square$

Now we let $\mathcal{H} = \bigcup \mathcal{G} \subset \mathcal{K}$. By the fact that $\mathcal{G}$ is centred, for any $K \in \mathcal{H}$,

$$K\backslash \bigcup\{U(H) : H \in \mathcal{H}, H \ne K)\} \ne \emptyset.$$

For each K we fix $x_K \in K$ and let $Y = \{x_K : K \in \mathcal{H}\} \subseteq X$. Note that since $\overline{Y} \in \mathcal{B}$ and $\overline{Y}\backslash Y = \overline{Y}\backslash \bigcup_{K \in \mathcal{H}} U(K) \in \mathcal{B}$, we have

$$Y = \overline{Y}\backslash(\overline{Y}\backslash Y) \in \mathcal{B}.$$

Note that since $x_K \notin U(H)$ for any distinct $H, K \in \mathcal{H}$, Y has the discrete topology. As a consequence the only compact sets in Y are the finite ones. Therefore

$$sup\{\mu(L) : L \subseteq Y, L \text{ } compact\} = 0. \qquad (4.1)$$

On the other hand, if U is open and $Y \subseteq U$, then for any $K \in \mathcal{H}$, $U\cap K \ne \emptyset$. Since all $K \in \mathcal{K}$ are μ-supporting, this means that U is an uncountable union of sets of positive measure. As a consequence $\mu(U) = \infty$, so we have

$$inf\{\mu(U) : U \supseteq Y, U \text{ } open\} = \infty. \qquad (4.2)$$

(4.1) and (4.2) suggest that μ cannot be both Radon and regular, contradicting our hypothesis. Therefore $\mathcal{K}$ must be countable. As a consequence $(X, \mathcal{B}, \mu)$ is σ-finite. $\square$

Chapter 5

Variations on the Souslin Hypothesis

5.1 The countable chain condition

Recall that a subset $\mathcal{X}$ of a partially ordered set $\mathcal{P}$ is *centred* if

$$\forall \mathcal{X}_0 \in [\mathcal{X}]^{<\omega} \ \exists p \in \mathcal{P} \ (\forall q \in \mathcal{X}_0 \ p \leq q).$$

We say that $\mathcal{P}$ is *σ-centred* if $\mathcal{P}$ can be decomposed into countably many centred subsets. Clearly, every σ-centred partially ordered set satisfies the countable chain condiction. The following result is a sort of converse to this.

Theorem 29. *Every ccc poset $\mathcal{P}$ of cardinality $< \mathfrak{m}$ is σ-centred.*

It could be shown that $\mathfrak{m}$ is the minimal cardinal θ with the property that every ccc poset of size $< \theta$ is σ-centred. So, Theorem 6 captures the full power of the Baire category cardinal $\mathfrak{m}$.

Before proving Theorem 29 we show the following:

Lemma. *If $\mathfrak{m} > \omega_1$ then the ccc is productive.*

Proof. Let $\mathcal{P}$ and $\mathcal{Q}$ be two ccc posets whose pruduct is not ccc and work towards a contradiction. Let (p_ξ, q_ξ), $\xi < \omega_1$, be a sequence of pairwise imcompatible elements of $\mathcal{P} \times \mathcal{Q}$. This means that for $\xi \neq \eta$, if p_ξ and p_η are compatible then q_ξ and q_η are incompatible.

For $\alpha < \omega_1$, let

$$\mathcal{D}_\alpha = \{p \in \mathcal{P} : \exists \xi \geq \alpha \ (p \leq p_\xi)\}.$$

Claim. *There is $r \in \mathcal{P}$ such that the set $\mathcal{D}_\alpha \cap \mathcal{P}(r)$ is dense in $\mathcal{P}(r)$ $(:= \{p \in \mathcal{P} : p \leq r\})$.*

Proof of claim. If not, then for every $s \in \mathcal{P}$ there is $r \leq s$ and $\alpha < \omega_1$ such that $\mathcal{P}(r) \cap \mathcal{D}_\alpha = \emptyset$. So we can get a maximal family $\mathcal{A}$ of pairwise incompatible elements of $\mathcal{P}$ with the property that for each $r \in \mathcal{A}$ there is $\alpha_r < \omega_1$ such that $\mathcal{P}(r) \cap \mathcal{D}_{\alpha_r} = \emptyset$.
Let
$$\beta = sup\{\alpha_r + 1 : r \in \mathcal{A}\}.$$
Then $\beta < \omega_1$ because $\mathcal{A}$ is countable. Find $r \in \mathcal{A}$ compatible with p_β and let $s \leq r, p_\beta$. Then $s \in \mathcal{P}(r) \cap \mathcal{D}_\beta \subseteq \mathcal{P}(r) \cap \mathcal{D}_{\alpha_r}$, a contradiction. $\square$

If $\mathfrak{m} > \omega_1$, there will be a filter $\mathcal{G}$ of some $\mathcal{P}(r)$ such that $\mathcal{G} \cap \mathcal{D}_\alpha \neq \emptyset$, for all $\alpha < \omega_1$. For each α, pick $r_\alpha \in \mathcal{G} \cap \mathcal{D}_\alpha$. Then, for each $\alpha < \omega_1$ there is $\xi_\alpha \geq \alpha$ such that $r_\alpha \leq p_{\xi_\alpha}$. It follows that p_{ξ_α} and p_{ξ_β} are compatible for all $\alpha < \beta < \omega_1$. Pick uncountable $\Gamma \subseteq \omega_1$ such that
$$(\forall \alpha < \beta \in \Gamma) \;\; \alpha \leq \xi_\alpha < \beta \leq \xi_\beta.$$
It follows that $(q_{\xi_\alpha})_{\alpha \in \Gamma}$, is a pairwise incompatible sequence of members of $\mathcal{Q}$, contradicting the assumption that $\mathcal{Q}$ is ccc. $\square$

Proof of Theorem 29. Let $\mathcal{P}$ be a ccc poset of size $< \mathfrak{m}$. Assume $\mathcal{P}$ has the maximum $\emptyset$. We may also assume that $\mathfrak{m} > \omega_1$. Consider
$$(\mathcal{P}^{\mathbb{N}})_{fin} = \{\hat{p} = (p_n)_{n \in \mathbb{N}} : \exists m \forall n \geq m \;\; p_n = \emptyset\}$$
ordered coordinate-wise. Note that $(\mathcal{P}^{\mathbb{N}})_{fin}$ satisfies the ccc iff all finite powers of $\mathcal{P}$ satisfy the ccc. So by the previous lemma $(\mathcal{P}^{\mathbb{N}})_{fin}$ is ccc.

For $r \in \mathcal{P}$, let $\mathcal{D}_r = \{\hat{p} \in (\mathcal{P}^{\mathbb{N}})_{fin} : (\exists n) \;\; p_n \leq r\}$.

Claim. $\mathcal{D}_r$ *is dense-open for all $r \in \mathcal{P}$.*

Proof of claim. Given $\bar{q} = (q_n)_{n \in \mathbb{N}}$ and $r \in \mathcal{P}$, let $k \in \mathbb{N}$ be such that $q_k = \emptyset$. Let $\hat{p} \in (\mathcal{P}^{\mathbb{N}})_{fin}$ be equal to $\hat{q}$ at all places except at $n = k$, where we put $p_k = r$. Then $\hat{p} \leq \hat{q}$ and $\hat{p} \in \mathcal{D}_r$. $\square$

Since $|\mathcal{P}| < \mathfrak{m}$ then there is a filter $\mathcal{G} \subseteq P^{\mathbb{N}})_{fin}$ such that $\mathcal{G} \cap \mathcal{D}_r \neq \emptyset$, for all $r \in \mathcal{P}$. For $k \in \mathbb{N}$, set
$$\mathcal{P}_k = \{p \in \mathcal{P} : (\exists \bar{q} = (q_n)_{n \in \mathbb{N}}) : q_k \leq p\}.$$
Then
$$\mathcal{P} = \bigcup_{k \in \mathbb{N}} \mathcal{P}_k$$
and each $\mathcal{P}_k$ is centred in $\mathcal{P}$. $\square$

5.2 The Souslin Hypothesis

Recall *Cantor's theorem* which says that every dense *separable* linearly ordered continuum is isomorphic to $[0, 1]$. The statement of the *Souslin Hypothesis* is:

Every dense linearly ordered continuum satisfying the countable chain condition is isomorphic to $[0, 1]$.

Any counterexample to the Souslin Hypothesis is called *Souslin continuum.*

Theorem 30 (Kurepa). *The countable chain condition of a Souslin continuum is not productive.*

Corollary. *If* $\mathfrak{m} > \omega$, *the Souslin hypothesis holds.*

Proof of Theorem 30. Let $\mathcal{K}$ be a Souslin continuum. By Cantor's theorem, we know that $\mathcal{K}$ is not separable. So we can build an increasing sequence D_ξ ($\xi < \omega_1$) of countable subsets of $\mathcal{K}$ such that for every maximal interval I of $\mathcal{K}$ missing D_ξ, $D_{\xi+1} \cap I \neq \emptyset$. To fix a notation, let $\mathcal{I}_\xi$ be the collection of all maximal intervals of $\mathcal{K}$ missing D_ξ. Moreover, we arrange $D_{\xi+1}$ in such a way that for every $I \in \mathcal{I}_\xi$ which is not separable, there exist disjoint $J_0(I), J_1(I) \subseteq I$ belonging to $\mathcal{I}_{\xi+1}$. Note that for each ξ there exists at least one $I_\xi \in \mathcal{I}_\xi$ which is not separable. Then

$$\{R_\xi := J_0(I_\xi) \times J_1(I_\xi) : \xi < \omega_1\}$$

is an uncountable family of pairwise disjoint open rectangles of $\mathcal{K}^2$.

To show this, fix $\xi < \eta$ and suppose $R_\xi \cap R_\eta \neq \emptyset$. Let $(x, y) \in R_\xi \cap R_\eta$. Then, $x \in J_0(I_\eta) \subsetneq J_0(I_\xi)$ (since for $\alpha < \beta$, I_β refines I_α). Similarly, $x \in I_\eta \subsetneq I_\xi$.

More precisely, $x \in J_0(I_\eta) \subsetneq I_\eta \subseteq J_0(I_\xi) \subsetneq I_\xi$. Similarly, $y \in J_1(I_\eta) \subsetneq I_\eta \subseteq J_1(I_\xi) \subsetneq I_\xi$. These two pictures are contradictory, since $J_0(I_\xi)$ and $J_1(I_\xi)$ are disjoint.

$\square$

Corollary. *There is no continuous map from a Souslin continuum into its square.*

$\square$

5.3   A selective ultrafilter from $\mathfrak{m} > \omega_1$

Definition. By a *coherent mapping* in this section we mean a mapping

$$a : [\omega_1]^2 \to \omega$$

such that:

1. $\{\xi < \alpha : a(\xi, \alpha) \neq a(\xi, \beta)\}$ is a finite set for all $\alpha < \beta < \omega_1$,

2. for every uncountable $X \subseteq \omega_1$ there is uncountable $X_0 \subseteq X$ such that $\Delta_a(\alpha, \beta) = \min\{\xi \leq \alpha : a(\xi, \alpha) \neq a(\xi, \beta)\} < \alpha$ for all $\alpha, \beta \in X_0$, $\alpha < \beta$.

Example. The function $\rho_1 : [\omega_1]^2 \to \omega$ of [32] is one example of such a mapping.

The following fact lists the crucial property of such an mapping and it will appear in Chapter 9 is a slightly bigger generality.

Theorem 31. *For every coherent mapping $a : [\omega_1]^2 \to \omega$, every positive integer n, and every uncountable family $\mathcal{F}$ of pairwise disjoint n-element subsets of ω_1 there is an uncountable $\mathcal{F}_0 \subseteq \mathcal{F}$ such that for all $s, t \in \mathcal{F}_0, s \neq t$ and all $i, j < n$, $\Delta_a(s(i), t(i)) = \Delta_a(s(j), t(j))$.*[1]

Corollary 32. *If $a : [\omega_1]^2 \to \omega$ is a coherent mapping then for every pair X and Y of uncountable subsets of ω_1 there is an uncountable subset Z of X such that $\Delta_a[Z] \subseteq \Delta_a[X] \cap \Delta_a[Y]$.*

The following definition will also appear in Chapter 9 in a slightly greater generality.

Definition. To any coherent mapping $a : [\omega_1]^2 \to \omega$ we attach the following family of subsets of ω_1,

$$\mathcal{U}_a = \{Y \subseteq \omega_1 : (\exists X \subseteq \omega_1) \, [X \text{ is uncountable and } \Delta_a[X] \subseteq Y]\}.$$

Theorem 33. *The family $\mathcal{U}_a$ is a uniform filter on ω_1, for every coherent mapping $a : [\omega_1]^2 \to \omega$.*

Proof. This is an immediate consequence of Corollary 32.   □

The following fact will also appear in Chapter 9.

Theorem 34. *If the countable chain condition is productive, and in particular, if $\mathfrak{m} > \omega_1$, then for every coherent $a : [\omega_1]^2 \to \omega$ the family $\mathcal{U}_a$ is a uniform ultrafilter on ω_1.*

---

[1]For an n-element set $t \subseteq \omega_1$, we let $t(i)$ $(i < n)$ be its increasing enumeration according to the natural order of ω_1.

Definition. An ultrafilter $\mathcal{V}$ on ω is *selective* if for every $h : \omega \to \omega$ there is $X \in \mathcal{V}$ such that h is either constant on X or is one-to-one on X.

Theorem 35. *Let* $a : [\omega_1]^2 \to \omega$ *be a coherent mapping, let* $f : \omega_1 \to \omega$, *and assume* $\mathfrak{m} > \omega_1$. *Then* $f[\mathcal{U}_a]$ *is a selective ultrafilter on* ω.

Proof. We may assume that $\mathcal{V} = f[\mathcal{U}_a]$ is non-principal since clearly principal ultrafilters are selective. To check the selectivity of $\mathcal{V}$, let $h : \omega \to \omega$ be a given mapping which is not constant on any set belonging to $\mathcal{V}$. We need to find $M \in \mathcal{V}$ such that $h \restriction M$ is one-to-one. This will be done by constructing an uncountable $X \subseteq \omega_1$ such that h is one-to-one on the set

$$M_f[X] = \{f(\Delta_a(\alpha, \beta)) : \alpha, \beta \in X, \alpha < \beta, \text{ and } \Delta_a(\alpha, \beta) < \alpha\}.$$

Clearly, any such a set $M_f[X]$ belongs to the ultrafilter $\mathcal{V}$. To this end, let $\mathcal{P}$ be the collection of all finite subsets p of ω_1 such that h is one to one on the set

$$M_f[p] = \{f(\Delta_a(\alpha, \beta)) : \alpha, \beta \in p, \alpha < \beta, \text{ and } \Delta_a(\alpha, \beta) < \alpha\}.$$

We consider $\mathcal{P}$ as a partially ordered set ordered by inclusion. Note that if $\mathcal{P}$ satisfies the countable chain condition our Baire category assumption $\mathfrak{m} > \omega_1$ would give us an uncountable $\mathcal{F} \subseteq \mathcal{P}$ such that $p \cup q \in \mathcal{P}$ for all $p, q \in \mathcal{F}$. Taking $X = \bigcup \mathcal{F}$ gives us an uncountable subset of ω_1 such that h is one-to-one on the corresponding set $M_f[X]$ which, as we know, belongs to the ultrafilter $\mathcal{V}$. To check the countable chain condition of $\mathcal{P}$ let $\mathcal{X}$ be an uncountable subset of $\mathcal{P}$. Refining $\mathcal{X}$ we may assume that $\mathcal{X}$ consists of n-element sets for some fixed positive integer n. By Theorem 31 and by the Δ-system lemma, we find uncountable $\mathcal{X}_0 \subseteq \mathcal{X}$ and an integer $n_0 < n$ such that for all $p, q \in \mathcal{X}_0$ such that $p \neq q$, we have that,

(1) $p(i) = q(i)$ for $i < n_0$ and $p(i) \neq q(i)$ for $n_0 \leq i < n$,

(2) $\Delta_a(p(i), q(i)) = \Delta_a(p(j), q(j))$ for $n_0 \leq i, j < n$.

For $p, q \in \mathcal{X}_0, p \neq q$, let $\Delta_a(p, q)$ denote the constant value of the sequence $\Delta_a(p(i), q(i))$ $(n_0 \leq i < n)$. Using the second property of the coherent mapping a and another Δ-system argument we arrive at an uncountable set $\mathcal{X}_1 \subseteq \mathcal{X}_0$ such that for all $p, q \in \mathcal{X}_1, p \neq q$,

$$\Delta_a[p \cup q] = \Delta_a[p] \cup \Delta_a[q] \cup \{\Delta_a(p, q)\}.$$

Find an integer k and uncountable $\mathcal{X}_2 \subseteq \mathcal{X}_1$ such that for all $p, q \in \mathcal{X}_2$,

$$f[\Delta_a(p)] = f[\Delta_a(p)] \subseteq \{0, 1, ..., k\}.$$

Now, find an integer ℓ such that $h(i) \leq \ell$ for all $i \leq k$. Let $Z = f^{-1}(\{0, 1, ..., \ell\})$. Then by our assumption about f the set Z does not belong to $\mathcal{U}_a$. By Corollary 32, there is an uncountable $\mathcal{X}_3 \subseteq \mathcal{X}_2$ such that

$$Z \cap \{\Delta_a(p, q) : p, q \in \mathcal{X}_3, p \neq q\} = \emptyset.$$

It follows that for arbitrary $p, q \in \mathcal{X}_3, p \neq q$ the function h is one-to-one on the set,

$$f[\Delta_a(p \cup q)] = f[\Delta_a[p]] \cup f[\Delta_a[q]] \cup \{f(\Delta_a(p,q))\}.$$

So, in particular $p \cup q \in \mathcal{P}$ for all $p, q \in \mathcal{X}_3$. This finishes the proof.   □

Corollary 36. *If* $\mathfrak{m} > \omega_1$, *there is a selective ultrafilter on* ω *which is* Σ_1*-definable in* $(H_{\omega_2}, \in)$

5.4   The countable chain condition versus the separability

Definition. A toplogical space X is *countably tight* (or *has countable tightness*) if for every $x \in X$ and $A \subseteq X$, if $x \in \overline{A}$ then $x \in \overline{A_0}$ for some countable $A_0 \subseteq A$.

Definition. A collection $\mathcal{U}$ of non-empty open sets is a π-*base* of a topological space X, if every non-empty $U \subseteq X$ contains a member of $\mathcal{U}$. The π-*weight* of X is the minimal cardinality of a π-base.

Theorem 37. *If* $\mathfrak{m} > \omega_1$ *then every compact countably tight ccc space is separable.*

Proof. Let K be a given compact countably tight ccc space and let π be its π-weight. Recursively, we construct a sequence (F_ξ, G_ξ) $(\xi < \pi)$ such that:

(1) F_ξ is a closed G_δ-subset of K with nonempty interior.

(2) G_ξ is an open F_σ-subset of K with $G_\xi \supseteq F_\xi$.

(3) G_ξ does not include any non-empty intersection of finitely many sets of the form
$$\{F_\gamma : \gamma < \xi\} \cup \{K \setminus G_\xi : \gamma < \xi\}.$$

Note that for each $\xi < \pi$ there is a non-empty open set G_ξ satisfying (3). Otherwise, for each finite intersection, $F_{AB} = (\bigcap_{\alpha \in A} F_\alpha) \cap (\bigcap_{\beta \in B} (K \setminus G_\alpha))$ is a closed G_δ set, so we can represent it as $F_{AB} = \bigcap_{n=0}^{\infty} \overline{V_{AB}^n}$, where V_{AB}^n is open, and observe that by compactness

$$\{V_{AB}^n : n \in \mathbb{N}, A, B \in [\xi]^{<\omega}, A < B\}$$

forms a π-base of X of size $|\xi| + \aleph_0$, contradicting the choice of π.

Claim. $\pi = \aleph_0$.

Proof of claim. Otherwise, by $\mathfrak{m} > \omega_1$ and the assumption that $\mathcal{P}_K = \{V \subseteq K : V \text{ is a nonempty open}\}$ is a ccc poset and $\{int(F_\xi) : \xi < \pi\}$ is an uncountable family of members of $\mathcal{P}_K$, by Theorem 29 there is an uncountable $\Gamma \subseteq \pi$ such that $\{int(F_\xi) : \xi \in \Gamma\}$ is centred. Going to a subset of Γ we may (and will) assume that $otp(\Gamma) = \omega_1$.

Subclaim. For every pair $A, B \subseteq \Gamma$ of finite sets such that $A < B$,

$$\left(\bigcap_{\alpha \in A} F_\alpha\right) \cap \left(\bigcap_{\beta \in B}(K \setminus G_\beta)\right) \neq \emptyset.$$

Proof of subclaim. By induction on $|B|$. If $B = \emptyset$ this follows from the fact that $\{int(F_\xi) : \xi \in \Gamma\}$ is centred. Suppose $|B| = k > 0$ and let $B_0 = B \setminus \{\xi\}$, where $\xi = max(B)$. By the inductive hypothesis, $F = (\bigcap_{\alpha \in A} F_\alpha) \cap (\bigcap_{\beta \in B_0}(K \setminus G_\beta)) \neq \emptyset$. By condition (3), $F \not\subseteq G_\xi$, so

$$\left(\bigcap_{\alpha \in A} F_\alpha\right) \cap \left(\bigcap_{\beta \in B}(K \setminus G_\beta)\right) = F \cap (K \setminus G_\xi) \neq \emptyset.$$

$\square$

This completes the proof of the claim. $\square$

And so the proof of Theorem 37. $\square$

Definition. A topological space X is T_5 if it is T_1 and for every pair A, B of subsets of X such that $\overline{A} \cap B = \overline{B} \cap A = \emptyset$ there exist disjoint open sets $U \supseteq A$ and $V \supseteq B$.

Note that every metric space is T_5.

Definition. A point x of some topological space X has *countable π-character* in X if there is a countable sequence $\{V_x^n\}_{n \geq 0}$ of open sets of X such that for every open $U \ni x$ there is n such that $U \supseteq V_x^n$. We say that X has *countable π-character* if every point of X has a countable π-character in X.

We shall need the following lemma which follows immediately from Theorem 42 below.

Lemma. *Every compact T_5 space has a point of countable π-character.*

Theorem 38. *If $\mathfrak{m} > \omega_1$ then every compact T_5 ccc space K is separable.*

Proof. By the previous lemma, the set

$$D = \{x \in K : x \text{ has countable } \pi\text{-character in } K\}$$

is dense in K, so it suffices to find a countable $D_0 \subseteq D$ dense in X. Suppose that $\overline{D_0} \neq K$ for all countable $D_0 \subseteq D$, and work for a contradiction.

For each $x \in D$ fix a countable family $\mathcal{U}_x$ of open subsets of K forming a π-base of x in K. By our assumption, we can build an increasing sequence D_ξ ($\xi < \omega_1$) of countable subsets of D such that, if for $\xi < \omega_1$ we let

$$\mathcal{U}_\xi = \bigcup_{x \in D} \mathcal{U}_x,$$

then we have the following condition satisfied:

(1) $D_\xi \subseteq D_{\xi+1} \nsubseteq \overline{D_\xi}$.

(2) $(\forall \mathcal{F} \in [\mathcal{U}_\xi]^{<\infty})[\bigcap \mathcal{F} \neq \emptyset \rightarrow \mathcal{F} \cap D_{\xi+1} \neq \emptyset]$.

Let

$$D_{\omega_1} = \bigcup_{\xi < \omega_1} D_\xi,$$

and let $Y = \overline{D_{\omega_1}}$. Let

$$\mathcal{U}_{\omega_1} = \bigcup_{\xi < \omega_1} \mathcal{U}_\xi.$$

Note that by (2), $\mathcal{U}_{\omega_1} \upharpoonright Y = \{U \cap Y : U \in \mathcal{U}_{\omega_1}\}$ is a π-base of Y and that Y is also a ccc space. By Theorem 29, the π-base $\mathcal{U}_{\omega_1} \upharpoonright Y$ is σ-centred, so we can write it as

$$\mathcal{U}_{\omega_1} \upharpoonright Y = \bigcup_{n \in \mathbb{N}} \mathcal{C}_n,$$

where each $\mathcal{C}_n$ is a centered family. By compactness, for each n we can choose a point

$$d_n \in \bigcap \{\bar{V} : V \in \mathcal{C}_n\}.$$

It follows that $\{d_n\}_{n \in \mathbb{N}}$ is a countable dense subset of Y. Pick $x \in Y$ such that $x \notin \overline{D_\xi}$ for all ξ. To see that such a point exists note that

$$M = \{n \in \mathbb{N} : (\forall \xi < \omega_1)\ x \notin \overline{D_\xi}\}$$

must be infinite, and in fact the closure $\overline{\{d_n : n \in M\}}$ must contain a set of the form $Y \setminus \overline{D_\xi}$ for some $\xi < \omega_1$. For each ξ choose an open neighborhood U_ξ of x in Y such that $\overline{U_\xi} \cap \overline{D_\xi} = \emptyset$ and moreover we have the following condition satisfied:

(3) $(\forall A \in [\omega_1]^\infty)(\exists \eta < \omega_1)\ \overline{U_\eta} \subseteq \bigcap_{\xi \in A} U_\xi$.

Let

$$F = \bigcap_{\xi \in \omega_1} U_\xi = \bigcap_{\xi \in \omega_1} \overline{U_\xi}.$$

Then F is a nonempty closed nowhere-dense subset of Y and U_ξ ($\xi \in \omega_1$) forms its neighborhood base in Y. Moreover,

$$(\forall \xi < \omega_1)\ F \cap \overline{D_\xi} \neq \emptyset.$$

Since $\overline{\{d_n : n \in M\}}$ contains $Y \setminus \overline{D_\eta}$ for some η, the family

$$M_\xi = \{n \in M : d_n \in U_\xi \setminus F\} \quad (\xi \in \omega_1)$$

is a family of infinite subsets of $\mathbb{N}$ such that

$$(\forall A \in [\omega_1]^{<\infty}) \quad \left| \bigcap_{\xi \in A} M_\xi \right| = \aleph_0.$$

By Theorem 16 we can find infinite $I \subseteq \mathbb{N}$ such that $I \setminus M_\xi$ is finite for all $\xi < \omega_1$. It follows that d_n $(n \in \mathbb{N})$ is a discrete subset of the open subspace $G = y \setminus F$. Since Y is T_5 we can find open sets $V_n \subseteq G$ $(n \in \mathbb{N}$ such that $d_n \in V_n$ for all $n \in I$ and such that V_n $(n \in \mathbb{N}$ is a discrete family of open subsets of G. Since D_{ω_1} is dense in Y, for each $n \in I$ we can choose $\xi_n < \omega_1$ such that $\xi_n < \eta$ for all $n \in I$. The subspace $\overline{D_\eta}$ of Y is compact so the sequence

$$V_n \cap \overline{D_\eta} \quad (n \in \mathbb{N})$$

of nonempty subsets of $\overline{D_\eta}$ must have an accumulation point $y \in \overline{D_\eta}$. By choice of V_n $(n \in \mathbb{N})$ any accumulation point of this family must belong to F, so we must have that $y \in F$, a contradiction.

□

Exercise. Prove directly that every compact T_5-space must have a point of countable π-character.

Exercise. Prove that every compact countably tight space has countable π-character.

Example. There is a compact ccc nonseparable space K in which every point has countable π-character. So in Theorem 37, we cannot replace countable tightness with the countable π-character.

Chapter 6

The S-spaces and the L-spaces

6.1 Hereditarily separable and hereditarily Lindelöf spaces

Definition. A topological space X is *hereditarily separable* if $\forall Y \subseteq X$ there exists a countable $Y_0 \subseteq Y$ such that $\overline{Y_0} = \overline{Y}$.

Definition. A topological space X is *hereditarily Lindelöf* if $\forall \mathcal{U} \subseteq \mathcal{OPEN}(X)$ there exists a countable $\mathcal{U}_0 \subseteq \mathcal{U}$ such that $\bigcup \mathcal{U}_0 = \bigcup \mathcal{U}$.

Notation. In case you have not already guessed, for a given topological space X, by $\mathcal{OPEN}(X)$ we denote the collection of all open subsets of X.

Now we state the S-space and L-space problems:

S-space problem When is a regular hereditarily separable space also hereditarily Lindelöf?

L-space problem When is a regular hereditarily Lindelöf space also hereditarily separable?

Now a few theorems concerning the S-space and L-space problems:

Theorem 39. *Assume $\mathfrak{m} > \omega_1$. Let X be a regular space with the property that every nonempty closed subspace F of X has a point x of countable π-character in F. Then every hereditarily Lindelöf subspace of X is hereditarily separable.*

Proof. At first note that every subspace of a hereditarily Lindelöf space is trivially hereditarily Lindelöf, so that it suffices to prove that every hereditarily Lindelöf subspace of X is separable. Now suppose $\exists Y \subseteq X$ such that Y is hereditarily Lindelöf and nonseparable. We will now proceed all the way to achieve a contradiction.

Since Y is nonseparable, by induction we can choose $\{y_\alpha : \alpha < \omega_1\} \subseteq Y$ such that $\forall \alpha < \omega_1$, $y_\alpha \in Y \setminus \overline{\{y_\beta : \beta < \alpha\}}$. Substitute Y by $\{y_\alpha : \alpha < \omega_1\}$ and then consider the obvious well-ordering $<_W$ (given by $y_\alpha \leq_W y_\beta$ iff $\alpha \leq \beta$) of Y to see that $\forall y \in Y$, $y \notin \overline{\{x \in Y : x <_W y\}}$. By regularity of X $\exists$ open $U_y \subseteq X$ such that

$$y \in U_y \subseteq \overline{U}_y \subseteq V_y := X \setminus \overline{\{x \in Y : x <_W y\}}.$$

Note that for each $x <_W y$ we have $x \notin \overline{U}_y$.

Now let $\mathcal{P}$ be the family of all finite $p \subseteq Y$ such that $x \notin U_y \ \forall x \neq y$ in p. We order $\mathcal{P}$ by reverse inclusion. If $\mathcal{P}$ satisfies the ccc then consider the family $\mathcal{P}_0 = \{\{y\} : y \in Y\} \subseteq \mathcal{P}$ of size ω_1. Since by our assumption $\mathfrak{m} > \omega_1$, by an earlier result proved in class we have that $\mathcal{P}_0$ is σ-centered. By regularity of ω_1 it follows that $\exists$ uncountable $\Lambda \subseteq \mathcal{P}_0$ such that Λ is centered in $\mathcal{P}$. Let $Z = \{y : \{y\} \in \Lambda\}$. Then $\{x, y\} \in \mathcal{P} \ \forall x, y \in Z$, which implies that $x \notin U_y \ \forall x \neq y \in Z$. But then there is no countable subcover of $\{U_y : y \in Z\} \implies Y$ is not hereditarily Lindelöf, which is a contradiction. It follows that $\mathcal{P}$ is *not* ccc.

Let P_ξ ($\xi < \omega_1$) be a sequence of pairwise incompatible conditions of $\mathcal{P}$. Applying the Δ-system lemma (taking a subsequence if necessary) we can assume that $\exists P \in \mathcal{P}$ such that

$$P_\xi \cap P_\eta = P \ \forall \xi \neq \eta \in \omega_1. \tag{P1}$$

Again, by regularity of ω_1, taking another subsequence if necessary we can assume that $\exists n < \omega$ such that

$$|P_\xi \setminus P| = n \ \forall \xi \in \omega_1. \tag{P2}$$

Finally, since by property (P1) above $(P_\xi \setminus P) \cap (P_\eta \setminus P) = \emptyset \ \forall \xi \neq \eta \in \omega_1$, we can assume that

$$\forall x \in P_\xi \setminus P, \ \forall y \in P_\eta \setminus P, \ x <_W y \text{ if } \xi < \eta. \tag{P3}$$

Now, $\forall \xi \in \omega_1$, by definition of $\mathcal{P}$ we trivially have that $P_\xi \setminus P \in \mathcal{P}$ and then pairwise incompatibility of our sequence and property (P1) above implies that $P_\xi \setminus P$ is incompatible with $P_\eta \setminus P \ \forall \eta \neq \xi$. So we may work with the sequence $P_\xi \setminus P$ ($\xi \in \omega_1$), or equivalently assume that $P = \emptyset$. Then, for each ξ let $\{P_\xi(0), \ldots, P_\xi(n-1)\} = P_\xi$ be a fixed enumeration of P_ξ.

Now consider $D_0 = \{P_\xi(0) : \xi \in \omega_1\}$. By hereditary Lindelöfness of Y $\exists \xi_0 \in \omega_1$ such that for every open $V \subseteq X$

$$V \cap \{P_\xi(0) : \xi_0 < \xi < \omega_1\} \neq \emptyset \implies |V \cap \{P_\xi(0) : \xi_0 < \xi < \omega_1\}| = \omega_1. \tag{P4}$$

For, otherwise for all $\xi \in \omega_1$, $\exists$ relatively open $V_\xi \subseteq Y$ such that $1 \leq |V_\xi \cap \{P_\eta(0) : \xi < \eta < \omega_1\}| \leq \omega$. But then $\{V_\xi : \xi < \omega_1\}$ defies the hereditary Lindelöfness of Y.

So, going to the subsequence $\{P_\xi(0) : \xi > \xi_0\}$ we may assume that for every open $V \subseteq X$

$$V \cap D_0 \neq \emptyset \implies |V \cap D_0| = \omega_1.$$

Let

$$Z_0 = \overline{D_0}.$$

By assumption on X, $\exists z_0 \in Z_0$ of countable π-character in Z_0. Then by the claim below $\{y \in Y : z_0 \in \overline{U}_y\}$ is countable.

Claim. *Let* $Y' \subseteq Y$, *and let* $y_0' \in \overline{Y'}$ *be of countable π-character in* $\overline{Y'}$. *Then* $\widetilde{Y}_0 = \{y \in Y : y_0' \in \overline{U}_y\}$ *is countable.*

Proof. Let $\mathcal{F}$ be a countable local π-base of y_0' in $\overline{Y'}$. Then $\forall y \in Y$ $y_0' \in \overline{U}_y \implies y_0' \in V_y$ $[\because \overline{U}_y \subseteq V_y] \implies \exists B_y \in \mathcal{F}$ such that $B_y \subseteq V_y$. Now, if $\widetilde{Y}_0$ is uncountable, then $|\mathcal{F}| \leq \omega \implies \exists B \in \mathcal{F}$ and uncountable $\widetilde{Y}_1 \subseteq \widetilde{Y}_0$ such that $B_y = B$ $\forall y \in \widetilde{Y}_1$. Pick any $y' \in B \cap Y'$. $\widetilde{Y}_1$ uncountable $\implies \exists y \in \widetilde{Y}_1$ such that $y' <_W y$. But then $y' \in B_y \subseteq V_y$, which is a contradiction to the definition of V_y. $\square$

Now, let $\mathcal{F}_0$ be a countable π-base of z_0 in Z_0. The above claim implies that $z_0 \notin \bigcup_{i<n} \overline{U}_{P_\xi(i)}$ for all but countably many ξ's. So for all but countably many $\xi \in \omega_1$ there exists $E_\xi \in \mathcal{F}_0$ such that $E_\xi \cap (\bigcup_{i<n} \overline{U}_{P_\xi(i)}) = \emptyset$. But then $\mathcal{F}_0$ countable $\implies \exists E_0 \in \mathcal{F}_0$ and uncountable $\Gamma_0 \subseteq \omega_1$ such that

$$E_0 \cap \left(\bigcup_{i<n} \overline{U}_{P_\xi(i)} \right) = \emptyset \ \forall \xi \in \Gamma_0.$$

Let $\Delta_0 = \{\xi \in \omega_1 : P_\xi(0) \in D_0 \cap E_0\}$. Since E_0 is a neighborhood of z_0 in $Z_0 = \overline{D_0}$, $E_0 \cap D_0 \neq \emptyset \implies$ by property (P4) above, Δ_0 is uncountable. But then $\forall \xi \in \Delta_0$, $\forall \eta \in \Gamma_0$ $P_\xi(0) \notin U_{P_\eta(i)}$ $\forall i < n$.

Now we consider

$$D_1 = \{P_\xi(1) : \xi \in \Delta_0\}$$
$$Z_1 = \overline{D_1}.$$

Similarly to the case of D_0, we may assume that if an open set intersects D_1 then the intersection is uncountable. And, in the same way, considering a point $z_1 \in Z_1$ of countable π-character in Z_1 we will get uncountable $\Gamma_1 \subseteq \Gamma_0$ and $\Delta_1 \subseteq \Delta_0$ such that for all $\xi \in \Delta_1$, $\forall \eta \in \Gamma_1$ $P_\xi(1) \notin U_{P_\eta(i)}$ $\forall i < n$.

Repeating this very process we will get uncountable $\Gamma_{n-1} \subseteq \Gamma_{n-2} \subseteq \ldots \subseteq \Gamma_0$, and $\Delta_{n-1} \subseteq \Delta_{n-2} \subseteq \ldots \subseteq \Delta_0$ such that $\forall \xi \in \Delta_{n-1}$, $\forall \eta \in \Gamma_{n-1}$ we have, $\forall i, j < n$, $P_\xi(j) \notin U_{P_\eta(i)}$. Now fix any $\eta \in \Gamma_{n-1}$, and then pick any $\xi \in \Delta_{n-1}$ such that $\xi > \eta$. Then $\forall i, j < n$, by the previous sentence $P_\xi(j) \notin U_{P_\eta(i)}$ and by property (P3) above and by definition of U_y's, $P_\eta(i) \notin U_{P_\xi(j)}$. But then $P_\xi \cup P_\eta$ is an extension of P_η and P_ξ in $\mathcal{P}$, and hence P_ξ and P_η are compatible, which is a contradiction. $\square$

Corollary 40. *If* $\mathfrak{m} > \omega_1$, *then any hereditarily Lindelöf subspace of a regular first countable space is separable.*

Proof. Any point in a first countable space X has countable π-character in X, and hence in any subspace of X. $\square$

Corollary 41. *Assume* $\mathfrak{m} > \omega_1$. *Let K be a compact space containing a hereditarily Lindelöf nonseparable subspace. Then K maps continuously onto* $[0,1]^{\omega_1}$.

Proof. By the theorem there exists a nonempty closed subspace F of K such that every point of F has uncountable π-character in F. But then we are done by the following well known result of Shapirovskii (see, [5]). $\square$

Theorem 42. *The following conditions are equivalent on a compact Hausdorff space X:*

(i) *X maps continuously onto the Tychonoff cube* $[0,1]^{\omega_1}$.

(ii) *X includes a closed nonempty subspace Y such that all points in Y have uncountable π-character in Y.*

6.2   Countable tightness and the S- and L-space problems

Definition. A topological space X is *countably tight* if $\forall Y \subseteq X$ and $y \in \overline{Y}$ $\exists$ countable $Y_0 \subseteq Y$ such that $y \in \overline{Y}_0$.

We shall need the following fact about this notion (see [5]).

Theorem. The product of two countably tight compact Hausdorff spaces is countably tight.

Definition. Let X be a topological space. $\{x_\xi : \xi < \theta\} \subseteq X$ is a *free sequence* for some cardinal θ if $\forall \gamma < \theta$

$$\overline{\{x_\xi : \xi < \gamma\}} \cap \overline{\{x_\xi : \xi \geq \gamma\}} = \emptyset.$$

We shall also need the following fact that connects this notion with the notion of countable tightness introduced above (see, for example, [5]).

Theorem. A compact Hausdorff space is countably tight iff it contains no uncountable free sequence.

Theorem 43. *Assume* $\mathfrak{m} > \omega_1$. *Let* X *be a regular space which contains no uncountable free sequence in any of its finite powers. Then a subspace* Y *of* X *is hereditarily separable iff* Y *is hereditarily Lindelöf.*

Proof. At first we will prove the "forward" direction. Suppose $Y \subseteq X$ is hereditarily separable but not hereditarily Lindelöf. Using $\mathfrak{m} > \omega_1$ we will produce an uncountable free sequence in X^k for some integer $k \leq 1$. So going to a subspace of Y we may assume that Y can be well ordered by $<_W$ in order type ω_1 such that $\forall y \in Y$, $\{x : x <_W y\}$ is relatively open in Y. (To see this, note that if Y is not hereditarily Lindelöf then there is a family $\mathcal{U}$ of relatively open subsets of Y such that $\forall$ countable $\mathcal{V}_0 \subseteq \mathcal{V}$, $\bigcup \mathcal{V}_0 \neq \bigcup \mathcal{V}$. But then by induction we can choose $V_\mu \in \mathcal{V}$ for all $\mu \in \omega_1$ such that $V_\mu \not\subseteq \bigcup_{\nu < \mu} V_\nu$. $\forall \mu \in \omega_1$ choose $y_\mu \in V_\mu \setminus \bigcup_{\nu < \mu} V_\nu$ and then replace Y by $\{y_\mu : \mu \in \omega_1\}$, with the ordering $<_W$ given by $y_\mu < y_\nu \iff \mu < \nu$.)

By regularity of X for each $y \in Y$ we can fix U_y and V_y open in X such that

1. $y \in U_y \subseteq \overline{U}_y \subseteq V_y$, and

2. $y <_W z \implies z \notin V_y$.

Let $\mathcal{P}$ be the collection of all finite subsets p of Y such that $z \notin U_y$ $\forall y \neq x \in p$. We order $\mathcal{P}$ by reverse inclusion.

Claim 43.1. $\mathcal{P}$ *does not satisfy the ccc.*

Proof. If $\mathcal{P}$ satisfies the ccc then consider the family $\mathcal{P}_0 = \{\{y\} : y \in Y\} \subseteq \mathcal{P}$ of size ω_1. Since by our assumption $\mathfrak{m} > \omega_1$, by an earlier result proved in class we have that $\mathcal{P}_0$ is σ-centered. By regularity of ω_1 it follows that $\exists$ uncountable $\Lambda \subseteq \mathcal{P}_0$ such that Λ is centered in $\mathcal{P}$. Let $Z = \{y : \{y\} \in \Lambda\}$. Then $\{x, y\} \in \mathcal{P}$ $\forall x, y \in Z$, which implies that $x \notin U_y$ $\forall x \neq y \in Z$. But then Z is not separable and so Y is not hereditarily separable, which is a contradiction. $\square$

Let P_ξ ($\xi < \omega_1$) be a sequence of pairwise incompatible conditions of $\mathcal{P}$. Applying the Δ-system lemma and then taking a subsequence if necessary (in the same way as in the proof of Theorem 39) we can assume that $\exists P \in \mathcal{P}$ and $k \in \omega$ such that

3. $P_\xi \cap P_\eta = P \,\, \forall \xi \neq \eta \in \omega_1$.

4. $|P_\xi \backslash P| = k \,\, \forall \xi \in \omega_1$.

5. $\forall x \in P_\xi \backslash P, \,\, \forall y \in P_\eta \backslash P, \,\, x <_W y$ if $\xi < \eta$.

By definition of $\mathcal{P}$ $P_\xi \backslash P \in \mathcal{P} \,\, \forall \xi \in \omega_1$, and $P_\xi \backslash P \perp P_\eta \backslash P \,\, \forall \eta \neq \xi$, we can replace P_ξ with $P_\xi \backslash P$ and assume $P = \emptyset$.

For $\eta \in \omega_1$ let

$$W_\eta^0 = \{(z_0, \ldots, z_{k-1}) \in X^k : \exists i, j < k \text{ such that } z_i \in U_{P_\eta(j)}\},$$
$$W_\eta^1 = \{(z_0, \ldots, z_{k-1}) \in X^k : \exists i, j < k \text{ such that } z_i \in V_{P_\eta(j)}\},$$

where $P_\eta = \{P_\eta(0), \ldots, P_\eta(k-1)\}$ is the $<_W$ increasing enumeration of P_η.

Now note that by (5) and (2) we have,

6. $\forall \xi > \eta \in \omega_1, (P_\xi(0), \ldots, P_\xi(k-1)) \notin W_\eta^1$.

Now fix $\xi < \eta < \omega_1$. $P_\xi \perp P_\eta \implies P_\xi \cup P_\eta \notin \mathcal{P} \implies$ either $P_\xi(i) \in U_{P_\eta(j)}$ for some $i, j < k$, or $P_\eta(i) \in U_{P_\xi(j)}$ for some $i, j < k$. But (6) excludes the latter possibility, so that we have the following:

7. $\forall \xi < \eta \in \omega_1, (P_\xi(0), \ldots, P_\xi(k-1)) \in W_\eta^0$.

But by definitions of U_η, V_η, W_η^0, W_η^1, we also have

8. $\forall \eta \in \omega_1, W_\eta^0 \subseteq \overline{W_\eta^0} \subseteq W_\eta^1$.

Fix $\eta \in \omega_1$. Since $X^k \backslash W_\eta^1$ is closed in X^k, (6), (7) and (8) imply that

$$\overline{\{(P_\xi(0), \ldots, P_\xi(k-1)) : \xi < \eta\}} \cap \overline{\{(P_\xi(0), \ldots, P_\xi(k-1)) : \xi > \eta\}} = \emptyset.$$

But then substituting P_ξ ($\xi \in \omega_1$) by a subsequence if necessary (for example can take P_{ξ_η} ($\eta \in \omega_1$) for any increasing subsequence ξ_η ($\eta \in \omega_1$) of ω_1 such that $\xi_\eta \geq \sup\{\xi_\beta : \beta < \eta\} + 2$ we will have the required "$\geq$" in the second set in the above identity, i.e. we will have

$$\overline{\{(P_\xi(0), \ldots, P_\xi(k-1)) : \xi < \eta\}} \cap \overline{\{(P_\xi(0), \ldots, P_\xi(k-1)) : \xi \geq \eta\}} = \emptyset,$$

so that $(P_\xi(0), \ldots, P_\xi(k-1))$ ($\xi < \eta$) is an uncountable free sequence in X^k. Hence we got the contradiction we wanted, and thus Y must be hereditarily Lindelöf. $\qquad\square$

Proving the other direction involves an argument totally symmetric to the one we just finished (so you can skip it, if you are not that persistent ...). Suppose $Y \subseteq X$ is hereditarily Lindelöf but not hereditarily separable. So going to a subspace of Y we may assume that Y can be well ordered by $<_W$ in order type ω_1 such that $\forall y \in Y$, $\{x : y <_W x\}$ is relatively open in Y. (cf. the 2nd paragraph in the proof of Theorem 39).

By regularity of X for each $y \in Y$ we can fix U_y and V_y open in X such that

1′. $y \in U_y \subseteq \overline{U_y} \subseteq V_y$, and

2′. $z <_W y \implies z \notin V_y$.

Let $\mathcal{P}$ be the collection of all finite subsets p of Y such that $z \notin U_y$ $\forall y \neq x \in p$. We order $\mathcal{P}$ by reverse inclusion.

Claim 43.1′. $\mathcal{P}$ *does not satisfy the ccc.*

Proof. Suppose the claim is false. Consider the family $\mathcal{P}_0 = \{\{y\} : y \in Y\} \subseteq \mathcal{P}$ of size ω_1. $\mathfrak{m} > \omega_1$ implies that there is uncountable $\Lambda \subseteq \mathcal{P}_0$ such that Λ is centered in $\mathcal{P}$. Let $Z = \{y : \{y\} \in \Lambda\}$. Then $\{x, y\} \in \mathcal{P} \ \forall x, y \in Z$, which implies that $x \notin U_y \ \forall x \neq y \in Z$. But then $\{U_y : y \in Z\}$ is a family of open sets which has no countable subcover and so Y is not hereditarily Lindelöf. Contradiction! $\square$

Let P_ξ ($\xi < \omega_1$) be a sequence of pairwise incompatible conditions of $\mathcal{P}$. Applying the Δ-system lemma and then taking a subsequence if necessary we can assume that $\exists P \in \mathcal{P}$ and $k \in \omega$ such that

3′. $P_\xi \cap P_\eta = P \ \forall \xi \neq \eta \in \omega_1$.

4′. $|P_\xi \backslash P| = k \ \forall \xi \in \omega_1$.

5′. $\forall x \in P_\xi \backslash P$, $\forall y \in P_\eta \backslash P$, $x <_W y$ if $\xi < \eta$.

By the same reasoning as in the previous case we can remove P and assume $P = \emptyset$.

For $\eta \in \omega_1$ let

$$W_\eta^0 = \{(z_0, \dots, z_{k-1}) \in X^k : \exists i, j < k \text{ such that } z_i \in U_{P_\eta(j)}\},$$
$$W_\eta^1 = \{(z_0, \dots, z_{k-1}) \in X^k : \exists i, j < k \text{ such that } z_i \in V_{P_\eta(j)}\},$$

where $P_\eta = \{P_\eta(0), \dots, P_\eta(k-1)\}$ is the $<_W$ increasing enumeration of P_η.

By (5′) and (2′) we have,

6′. $\forall \xi < \eta \in \omega_1$, $(P_\xi(0), \dots, P_\xi(k-1)) \notin W_\eta^1$.

Now fix $\eta < \xi < \omega_1$. $P_\xi \perp P_\eta \implies P_\xi \cup P_\eta \notin \mathcal{P} \implies$ either $P_\xi(i) \in U_{P_\eta(j)}$ for some $i, j < k$, or $P_\eta(i) \in U_{P_\xi(j)}$ for some $i, j < k$. But (6') excludes the latter possibility, so that we have the following:

7'. $\forall \xi > \eta \in \omega_1$, $(P_\xi(0), \ldots, P_\xi(k-1)) \in W_\eta^0$.

But by definitions of U_η, V_η, W_η^0, W_η^1, we also have

8'. $\forall \eta \in \omega_1$, $W_\eta^0 \subseteq \overline{W_\eta^0} \subseteq W_\eta^1$.

Fix $\eta \in \omega_1$. Since $X^k \setminus W_\eta^1$ is closed in X^k, (6'), (7') and (8') imply that

$$\overline{\{(P_\xi(0), \ldots, P_\xi(k-1)) : \xi < \eta\}} \cap \overline{\{(P_\xi(0), \ldots, P_\xi(k-1)) : \xi > \eta\}} = \emptyset.$$

Again, by substituting P_ξ ($\xi \in \omega_1$) by a subsequence if necessary we will have the required "$\geq$" in the second set in the above identity and which will give us an uncountable free sequence in X^k, which is a gross contradiction to our assumption. Thus Y must be hereditarily separable. $\qquad\square$

Corollary 44. *The following conditions on a regular space X are equivalent if $\mathfrak{m} > \omega_1$:*

(i) All finite powers of X are hereditarily separable.

(ii) All finite powers of X are hereditarily Lindelöf.

(iii) No finite power of X contains an uncountable discrete subspace.

Proof. Assume $\mathfrak{m} > \omega_1$. At first we prove (i) $\implies$ (ii). So let X be a regular space and suppose all finite powers of X are hereditarily separable. Fix an integer $k \geq 1$. Let $\{x_\nu : \nu < \omega_1\} \subseteq X^{mk}$ for some integer $m \geq 1$. Since X^{mk} is hereditarily separable, $\exists \nu_n \in \omega_1$ ($n \in \omega$) such that

$$\overline{\{x_\nu : \nu < \omega_1\}} = \overline{\{x_{\nu_n} : n < \omega\}}.$$

Let $\mu = \sup\{x_{\nu_n} : n < \omega\} + 1 < \omega_1$. Then it follows that

$$\overline{\{x_\nu : \nu \geq \mu\}} \subseteq \overline{\{x_\nu : \nu < \mu\}},$$

so that x_ν ($\nu < \omega_1$) is not a free sequence. Since m was arbitrary and since any uncountable free sequence contains a free sequence of size ω_1, it follows that no finite power of X^k contains an uncountable free sequence. But then by Theorem 43 X^k is hereditarily Lindelöf. Since k was arbitrary it follows that all finite powers of X are hereditarily Lindelöf. $\qquad\square$

Now we will deal with (ii) $\implies$ (iii). This is easy. Let X be a regular space such that all finite powers of X are hereditarily Lindelöf. But then *no* finite power X^k of X can contain an uncountable discrete subspace Y, for otherwise for each $y \in Y$, $\exists$ open $U_y \subseteq X^k$ such that $U_y \cap Y \setminus \{y\} = \emptyset \implies \{U_y : y \in Y\}$ is a collection of open subsets of X^k that defies the hereditary Lindelöfness of X^k. $\square$

Now the last one: (iii) $\implies$ (i). Suppose this is not true, and assume that no finite power of the regular space X' contains an uncountable discrete subspace and that $\exists n \geq 1$ such that $(X')^n$ is not hereditarily separable. Now consider the 2nd part of the proof of Theorem 43, where we took a regular space X and a subspace Y which is hereditarily Lindelöf but not hereditarily separable, and produced an uncountable free sequence in X^k for some $k \geq 1$. The only place where the assumption of Y being hereditarily Lindelöf was used was in Claim 43.1', to prove that there is no discrete subset Z of Y of size ω_1. But look here, we already have this property for $(X')^n$, since by our assumption on X, $(X')^n$ contains no uncountable discrete subset! So we can proceed exactly in the same way to get the existence of an uncountable free sequence $\{a_\xi : \xi < \omega_1\}$ in $(X')^{nk}$, for some $k \geq 1$. But then note that, for any $\xi < \omega_1$,

1. $\overline{\{a_\eta : \eta < \xi\}} \cap \overline{\{a_\eta : \eta \geq \xi\}} = \emptyset \implies a_\xi \notin \overline{\{a_\eta : \eta < \xi\}}$, and that

2. $\overline{\{a_\eta : \eta < \xi + 1\}} \cap \overline{\{a_\eta : \eta \geq \xi + 1\}} = \emptyset \implies a_\xi \notin \overline{\{a_\eta : \eta \geq \xi + 1\}}$.

Since $(X')^{nk}$ is regular it follows that $\exists$ open neighborhoods U_ξ and V_ξ of a_ξ such that

$$U_\xi \cap \overline{\{a_\eta : \eta < \xi\}} = \emptyset,$$
$$V_\xi \cap \overline{\{a_\eta : \eta > \xi\}} = \emptyset,$$

so that $U_\xi \cap V_\xi$ is an open neighborhood of a_ξ that does not contain any a_η for $\eta \neq \xi$. But then it follows that $\{a_\xi : \xi < \omega_1\}$ is a discrete subspace of $(X')^{nk}$ of size ω_1, which is a contradiction. $\square$

Corollary 45. *Assume* $\mathfrak{m} > \omega_1$. *Let K be a compact Hausdorff countably tight space. Then a subspace of K is hereditarily separable iff it is hereditarily Lindelöf.*

Proof. Recall that a compact space is countably tight if and only if it contains no uncountable free sequence and that product of two compact countably tight spaces is countably tight. It follows that no finite power of K contains an uncountable free sequence. Then the result follows from Theorem 43. $\square$

Theorem 46. *Let X be a regular space with no uncountable free sequence in any of its finite powers. Suppose Y is a subspace of X of cardinality less than $\mathfrak{m}$, and also that (U_y, V_y) $(y \in Y)$ is a sequence of pairs of open subsets of X such that $\overline{U}_y \subseteq V_y$ for all $y \in Y$, and one of the following conditions holds:*

(a) $V_y \cap Y$ is countable $\forall y \in Y$, or

(b) $\{y \in Y : z \in V_y\}$ is countable $\forall z \in Y$.

Then Y can be decompose as $Y = \bigcup_{n=0}^{\infty} Y_n$, so that

$$(\forall n \in \omega)(\forall y \neq z \in Y_n) \quad z \notin U_y.$$

Proof. Let $\mathcal{P}$ be the collection of all finite subset p of Y such that $z \notin U_y$ $\forall y \neq z \in p$. We order $\mathcal{P}$ by reverse inclusion. Since $|\mathcal{P}| < \mathfrak{m}$, the result follows if we can prove that $\mathcal{P}$ satisfies the ccc, for then by a result proved earlier, we will have that $\mathcal{P}$ is σ-centered, which is precisely the statement of the theorem.

So suppose $\mathcal{P}$ does not satisfy the ccc, and let P_ξ ($\xi < \omega_1$) be a sequence of pairwise incompatible conditions of $\mathcal{P}$. Applying the Δ-system lemma and then taking a subsequence if necessary we can assume that $\exists P \in \mathcal{P}$ and $k \in \omega$ such that

$$P_\xi \cap P_\eta = P \; \forall \xi \neq \eta \in \omega_1. \tag{6.1}$$

$$|P_\xi \backslash P| = k \; \forall \xi \in \omega_1. \tag{6.2}$$

Since by definition of $\mathcal{P}$, $P_\xi \backslash P \in \mathcal{P}$ for all $\xi \in \omega_1$, we can remove P and assume $P = \emptyset$. Then, if (a) holds, then since our sequence is uncountable, taking a subsequene if necessary we can assume that

$$P_\xi \cap V_y = \emptyset \quad \forall y \in P_\eta, \; \forall \xi > \eta. \tag{6.3a}$$

Similarly, (b) would let us assume that

$$P_\eta \cap V_y = \emptyset \quad \forall y \in P_\xi, \; \forall \xi > \eta. \tag{6.3b}$$

But then, as in the proof of Theorem 43, for $\eta \in \omega_1$ let

$$W_\eta^0 = \{(z_0, \ldots, z_{k-1}) \in X^k : \exists i, j < k \text{ such that } z_i \in U_{P_\eta(j)}\},$$

$$W_\eta^1 = \{(z_0, \ldots, z_{k-1}) \in X^k : \exists i, j < k \text{ such that } z_i \in V_{P_\eta(j)}\},$$

where $P_\eta = \{P_\eta(0), \ldots, P_\eta(k-1)\}$ is a fixed enumeration of P_η for all $\eta \in \omega_1$.

By definitions of U_η, V_η, W_η^0, W_η^1, we wiil then have

$$\forall \eta \in \omega_1, \; W_\eta^0 \subseteq \overline{W_\eta^0} \subseteq W_\eta^1. \tag{6.4}$$

Now consider the sequence $(P_\xi(0), \ldots, P_\xi(k-1))$ $(\xi < \omega_1)$ in X^k. In the same way as in the proof of the Theorem 43, the incompatibility of $\{P_\xi : \xi < \omega_1\}$, (6.4), and either (6.3a) or (6.3b) would give that

$$\overline{\{(P_\xi(0), \ldots, P_\xi(k-1)) : \xi < \eta\}} \cap \overline{\{(P_\xi(0), \ldots, P_\xi(k-1)) : \xi \geq \eta\}} = \emptyset.$$

Then we have a free sequence in X^k, contrary to our assumption on X. It follows that $\mathcal{P}$ *is* ccc, and we are done. $\square$

Chapter 7

The Side-condition Method

7.1 Elementary submodels as side conditions

Many of the proofs in this and subsequent lectures will involve the use of *elementary submodels as side conditions* in partial orders. For general information about elementary submodels, including some of their applications in set theory, see [11]. Here we will include some important facts about elementary submodels that will be used in this lectures.

Notation. Recall that for any cardinal θ, we write

$$H_\theta = \{x : |TC(x)| < \theta\},$$

that is, the family of all sets hereditarily of cardinality less than θ. Then $(H_\theta, \in)$ is a model of all the axioms of ZFC except the instances of the power-set axiom or the axiom of replacement when θ is singular. We will be considering only standard models of sufficiently large fragments of ZFC, that is models of the form $\langle M, \in \rangle$, where M is a set, and $\in$ denotes the membership relation on elements of M. We will usually use just the name of the set, such as M, to refer to the model $(M, \in)$, so that the $\in$ relation will always be implied. For models A and B, we write $A \prec B$ to mean that A is an elementary submodel of B (which of course should really be written in full as $\langle A, \in \rangle \prec \langle B, \in \rangle$).

Although the $\in$ relation is not generally transitive in the class of elementary submodels of a given $(H_\theta, \in)$, we will often want to refer to "$\in$-chains of countable elementary submodels" of some transitive model. The following fact allows us to do this:

Lemma. *Suppose θ is a regular uncountable cardinal, and suppose A, B, and C are countable elementary submodels of H_θ. Then the following facts are true:*

1. *If $A \in B$ and $B \in C$ then $A \in C$ (so $\in$ is a strict partial order relation on countable elementary submodels of the transitive model H_θ; this allows us to talk about "$\in$-chains of countable elementary submodels", sometimes known as* elementary $\in$-chains*); and*

2. *If $A \in B$ then $A \prec B$ and so an 'elementary $\in$-chain' is in fact a chain under $\prec$.*

Lemma. *Suppose θ is an uncountable regular cardinal, suppose $X \in H_\theta$ is countable, and suppose $S \subseteq \omega_1$ is a stationary set. Then there exists a countable elementary submodel $M \prec H_\theta$ such that $X \subseteq M$ and $\omega_1 \cap M \in S$.*

Lemma. *Suppose $\kappa < \lambda$ are cardinals and $M \prec H_\lambda$ and $\kappa \in M$. Then we have*

$$M \cap H_\kappa \prec H_\kappa.$$

Definition (The $\in$-collapse). Fix a sufficiently large regular cardinal θ. The $\in$-*collapse of H_θ* is the poset $\mathcal{P}$ of all finite $\in$-chains p of countable elementary submodels of H_θ ordered by the inclusion.

Theorem 47. *The $\in$-collapse of an arbitrary regular H_θ is stationary preserving and, in fact, canonically proper.*

Proof. Let $C(p)$ ($p \in \mathcal{P}$) be a $\mathcal{P}$-club. Let $\bar{p} \in \mathcal{P}$ and $E \in \mathrm{stat}(\omega_1)$ be given. Choose a countable elementary submodel M of $(H_{(2^\theta)^+}, \in)$ containing all these objects such that $\delta = M \cap \omega_1 \in E$. Let

$$p = p \cup \{M \cap H_\theta\}.$$

Clearly p belongs to $\mathcal{P}$. Choose q extending p such that $C(q) \setminus \delta \neq \emptyset$. We claim that $\delta \in C(q)$. Otherwise, we can find $r \leq q$ and $\gamma < \delta$ such that

$$(\forall s \leq r) C(s) \cap (\gamma, \delta] = \emptyset.$$

Let $\bar{r}_0 = r \cap M$ and choose $\bar{\delta} < \delta$ above γ and

$$\sup\{N \cap \omega_1 : N \in \bar{q}\}.$$

Using the elementaristy of M, we can find $\bar{r} \in \mathcal{P} \cap M$ end-extending $\bar{r}_0$ such that $C(\bar{r}) \setminus \bar{\delta} \neq \emptyset$. However, note that $\bar{r}$ and r are compatible, their union $\bar{r} \cup r$ being the common extension. But this is a contradiction since

$$C(\bar{r} \cup r) \cap (\gamma, \delta] \neq \emptyset.$$

$\square$

In numerous examples seen from now on in these lectures, we are going to see that the elements of the $\in$-collapse serve very well as the side conditions to posets whose 'working parts' are introducing the objects that we really want. The finite $\in$-chains of countable elementary submodels are there to facilitate the proofs that the resulting posets are stationary preserving or proper.

The $\in$-collapse itself is an interesting poset. It clearly collapses H_θ to be of cardinality $\aleph_1$ since it represents $H(\theta)$ as the union of an ω_1-chain ordered by $\in$ of countable elementary submodels of H_θ. We have already indicated that this is a proper poset, but if we restrict the countable submodels to belong to a fixed set $S \subseteq [H_\theta]^\omega$ then we may obtain weaker properties. Call this variation of the $\in$-collapse by $\mathcal{P}(E)$. Then we obtain the following variations of Theorem 47.

Theorem 48. *If S is projectively stationary (see Chapter 17) then $\mathcal{P}(E)$ is stationary preserving.*

Theorem 49. *If S is semi-closed and unbounded (see definition below) then $\mathcal{P}(E)$ is semi-proper.*

Definition. A set $S \subseteq [H_\theta]^\omega$ is *semi-closed and unbounded* if there is a closed and unbounded set $C \subseteq [H_\theta]^\omega$ such that for every $N \in C$ there is $M \in S$ such that $M \supseteq N$ but $M \cap \omega_1 = N \cap \omega_1$.

The $\in$ collapse can of course be applied to subsets of $[\kappa]^\omega$. For example, since $[\omega_2]^\omega$ has a projectively stationary subset of cardinality $\aleph_2$, we get the following result.

Theorem 50. *There is a stationary preserving poset of cardinality $\aleph_2$ which collapses $\aleph_2$ to $\aleph_1$.*

It turns out that such a poset cannot be found in the more restrictive classes the following result shows.

Theorem 51. *If a semi-proper poset collapses $\aleph_2$ then it also collapses the continuum.*

Corollary 52. *If $\mathfrak{c} > \aleph_2$ then there is no semi-proper poset collapsing $\aleph_2$ but preserving all other cardinals.*

7.2 Open graph axiom

There are various equivalent formulations of the Open Graph Axiom. To introduce this axiom we need the following piece of notation

Notation. For any set X, we define:

$$X^{[2]} = X^2 \setminus \{(x, x) : x \in X\}.$$

If X is a topological space, we give $X^{[2]}$ the topology inherited as a subspace of X^2, where X^2 is given the product topology. Note that if X is Hausdorff, then its diagonal is closed, and so $X^{[2]}$ is open, and so a subset $K \subseteq X^{[2]}$ is open in $X^{[2]}$ iff it is an open subset of X^2.

Definition. For any set X, we say that $R \subseteq X^{[2]}$ is *symmetric* if

$$(\forall x, y \in X)\left[(x, y) \in R \iff (y, x) \in R\right].$$

Notation. For any set X, any $R \subseteq X^{[2]}$, and any $x \in X$, we define the set

$$R(x) = \{z \in X : (x, z) \in K\}.$$

Definition. The *Open Graph Axiom (OGA)* or *Todorcevic Axiom (TA)* refers to the following statement:

Let X be a separable metric space. (More generally, we can let X be a second-countable Hausdorff space.) Then for every *open graph* on X, i.e., a structure of the form $\mathcal{G} = (X, R)$ for some open symmetric $R \subseteq X^{[2]}$, either:

1. $\mathcal{G}$ has an uncountable *complete subgraph*, i.e., there is an uncountable $Y \subseteq X$ such that $Y^{[2]} \subseteq R$, or

2. $\mathcal{G}$ is *countably chromatic* i.e., there is a decomposition

$$X = \bigcup_{n=1}^{\infty} X_n$$

such that for all n, we have $X_n^{[2]} \cap R = \emptyset$.

It turns out that this dichotomy is false for *closed* graphs.

Fact. *For $X = \mathbb{N}^{\mathbb{N}}$, there is a closed symmetric $K \subseteq X^{[2]}$ such that neither alternative of the conclusion of OGA holds for K.*

It is *not* a theorem that $\mathfrak{m} > \omega_1 \Rightarrow$ OGA. However, by starting with a stronger hypothesis, we do have the following theorem:

Theorem 53. *If $\mathfrak{mm} > \omega_1$ then OGA is true.*

Proof. More precisely, we shall show that given an open graph $\mathcal{G} = (X, R)$ on a separable metric space X such that $\mathrm{Chr}(\mathcal{G}) > \aleph_0$ there is a stationary set preserving poset which forces that $\mathcal{G}$ has an uncountable complete subgraph, i.e., $\mathcal{K}_{\aleph_1} \leq \mathcal{G}$.

Let $\mathcal{P}$ be the collection of all pairs $p = \langle H_p, \mathcal{N}_p \rangle$ with the following properties:

1. H_p is a finite *complete* set of vertices of $\mathcal{G}$, i.e., H is a finite subset of X such that for every $x \neq y \in H_p$ the pair $\{x, y\}$ is an edge of $\mathcal{G}$,

2. $\mathcal{N}_p$ is a finite $\in$-chain of countable elementary submodels N of $(H_{\mathfrak{c}^+}, \in)$,

3. $\mathcal{N}_p$ separates H_p, that is,

$$(\forall x \neq y \in H_p)\,(\exists N \in \mathcal{N}_p)\,|N \cap \{x, y\}| = 1,$$

4. $(\forall N \in \mathcal{N}_p)(\forall x \in H_p \setminus N)(\forall Y \in N)[x \in Y \rightarrow Y^2 \cap R \neq \emptyset]$.

We set $p \leq q$ if and only if

$$H_p \supseteq H_q \text{ and } \mathcal{N}_p \supseteq \mathcal{N}_q.$$

The proof of the Theorem follows easily from the following Claim.

Claim 1. $\mathcal{P}$ is a stationary-set preserving poset.

Let $(C(p) : p \in \mathcal{P})$ be a given $\mathcal{P}$-club, let $\bar{p}$ be a given condition of $\bar{p}$ and let E be a given stationary subset of ω_1. Find a countable

$$\bar{p}, E, \mathcal{G} \in M \prec \left(H_{(2^{\mathfrak{c}})^+}, \in \right)$$

such that $\delta = M \cap \omega_1 \in E$. Let

$$p = \left(H_{\bar{p}}, \mathcal{N}_{\bar{p}} \cup \{M \cap H_{\mathfrak{c}^+}\} \right).$$

Choose $q \leq p$ such that $C(q) \setminus \delta \neq \emptyset$. We claim that $\delta \in C(q)$. Otherwise, we can find $\gamma < \delta$ and $r \leq q$ such that

$$(\forall s \leq r)C(s) \cap (\gamma, \delta] = \emptyset.$$

We assume γ is above all the ordinals of the form $N \cap \omega_1$ for $N \in \mathcal{N}_r \cap M$.

Let $\mathcal{X}$ be the collection of all condition X of $\mathcal{P}$ that end extend $r \cap M$ and that are isomorphic with r over the parameters γ, $r \cap M$. We will find $\bar{r} \in \mathcal{X} \cap M$ compatible with r. This leads us to the desired contradiction, since $C(\bar{r}) \setminus (\gamma + 1) \neq \emptyset$ which by the elementarity of M means that $C(\bar{r})$ intersects the interval $(\gamma, \delta]$ so any common extension of r and r will also have this property contradicting the choice of r and γ.

Clearly $r \in \mathcal{D}$. Let

$$n = |H_r \setminus M|.$$

Let

$$x^r = \left\langle x_0^r, x_1^r, \ldots, x_{n-1}^r \right\rangle,$$

where

$$H_r \setminus M = \left\{ x_0^r, x_1^r, \ldots, x_{n-1}^r \right\},$$

with indexing chosen according to the $\in$-ordering of $\mathcal{N}_r$:

$$(\forall i < n - 1)\,(\exists N \in \mathcal{N}_r)\,[x_i \in N \ \& \ x_{i+1} \notin N].$$

Let
$$\mathcal{F} = \{x \in X^n : \exists s \le r \cap M\,[s \in \mathcal{D} \;\&\; x^s = x]\}.$$

Then $\mathcal{F} \in M$ and the family $\mathcal{F}$ is large in the following precise sense.

Claim 2. The family $\mathcal{F}$ of n-tuples is *large* relative to the co-ideal
$$Q = \{Y \subseteq X : \mathrm{Chr}\,(\mathcal{G} \restriction Y) > \aleph_0\},$$

or in other words the following formula is true
$$Qx_0 Qx_1 \ldots Qx_{n-1} \langle x_0, x_1, \ldots, x_{n-1}\rangle \in \mathcal{F}.[1]$$

Note that large families of n-tuples of elements of X form a nontrivial σ-complete co-ideal of subsets of X^n.

Claim 3. For almost all $x \in \mathcal{F}$, in the sense that the set of exceptions does not contain a large family, there is $y \in \mathcal{F}$ such that
$$\forall i < n\,[(x_i, y_i) \in R].$$

Proof. Induction on n. Let
$$\mathcal{H}_0 = \{x \in X^{n-1} : \mathrm{Chr}\,(\mathcal{G} \restriction Y_x) > \aleph_0 \text{ for } Y_x = \{a \in X : x^\frown a\} \in \mathcal{F}\}.$$

Then from the assumption that $\mathcal{F}$ is a large subset of X^n, we conclude that $\mathcal{H}_0$ is a large subset of X^{n-1}. Fix a countable base $\mathcal{B} \in M$ for X. Since for $x \in \mathcal{H}_0$ we have that in particular $Y_x^2 \cap R \ne \emptyset$, and since R is open, we can fix $U_x, V_x \in \mathcal{B}$ such that

1. $U_x \cap V_x = \emptyset$,
2. $U_x \cap Y_x \ne \emptyset \ne V_x \cap Y_x$,
3. $U_x \times V_x \subseteq R$.

Find a large $\mathcal{H}_1 \subseteq \mathcal{H}_0$ and $U, V \in \mathcal{B}$ such that
$$(\forall x \in \mathcal{H}_1)\,[U_x = U \;\&\; V_x = V].$$

By the inductive hypothesis there exist $x, y \in \mathcal{H}_1$ such that
$$(\forall i < n-1)\,(x_i, y_i) \in R.$$

Find $a \in U$ such that $x^\frown a \in \mathcal{F}_0$. Find $b \in V$ such that $y^\frown b \in \mathcal{F}_0$. Then $x^\frown a$ and $y^\frown b$ are as required. $\square$

---

[1] Recall that for a formula $\varphi(v)$ with one free variable v ranging over the set X, the formula $Qx\varphi(v)$ is interpreted as saying that $\{x \in X : \varphi(x)\} \in Q$.

Note that this finishes the proof of Claim 1, and therefore the proof of Theorem 53. $\square$

The following is an important open problem about OGA.

Problem 7.2.1. Does OGA imply $2^{\aleph_0} = \aleph_2$?

Since OGA implies $\mathfrak{b} = \aleph_2$ this is equivalent to the following question.

Problem 7.2.2. Does OGA imply $\mathfrak{b} = \mathfrak{c}$.

Exercise. Show that OGA implies $\mathfrak{b} = \aleph_2$.

Exercise. Show that the dual of OGA is false, where by dual we mean the statement that every *closed* graph on a separable metric space is either countably chromatic or it contains an uncountable complete subgraph.

Chapter 8

Ideal Dichotomies

We will prove in this Chapter that $\mathfrak{mm} > \omega_1$ implies two important statements involving ideals of a given set S, which, we will see imply some other important combinatorial properties of ω_1.

8.1 Small ideal dichotomy

We will assume in what follows that all the ideals we are considering are proper ideals on a given set S. In fact we shall be primarily interested in ideals consisting only of countable subsets of S.

Definition. Let $\mathcal{I}$ be an ideal of countable subsets of some set S. We say that $\mathcal{I}$ is $< \kappa$- generated if there exists a family $\mathcal{F} \subseteq \mathcal{P}(S)$ such that

1. $|\mathcal{F}| < \kappa$

2. $(\forall A \in \mathcal{I})\,(\exists F \in \mathcal{F})(A \subseteq^* F)$.

We say that $\mathcal{F}$ is a generating family for $\mathcal{I}$.

Definition. Let X a nonempty set, $\mathcal{F} \subseteq \mathcal{P}(X)$ and $T \subseteq X$. We say that T is orthogonal to $\mathcal{F}$ iff $T \cap F \in Fin$ for every $F \in \mathcal{F}$. Notice that if $\mathcal{F}$ generates an ideal $\mathcal{I}$ and T is orthogonal to $\mathcal{F}$, then T is orthogonal to $\mathcal{I}$.

Theorem 54. *Let $\mathcal{I}$ be an ideal of countable subsets in some set S. Suppose $\mathcal{I}$ is $< \mathfrak{mm}$-generated. Then either.*

1. *There exists an uncountable $T \subseteq S$ orthogonal to $\mathcal{I}$, or*

2. *There is a decomposition* $S = \bigcup\limits_{n=0}^{\infty} S_n$ *such that* $[S_n]^{\leq\omega} \subseteq \mathcal{I}$ *for all* $n \in \mathbb{N}$.

The conclusion of this theorem for $\aleph_1$-generated ideals will be called the *Small Ideal Dichotomy* and denoted **SID**.

Proof. Assuming that (2) does not hold. Note that this in particular implies that $\mathrm{mm} > \omega_1$. We will use this Baire category assumption to produce $T \subseteq S$ orthogonal to $\mathcal{F}$, where $\mathcal{F}$ is the $< \mathrm{mm}$ generating family of $\mathcal{I}$, proving in this way the theorem.

Let $\mathcal{P}$ be the collection of $p = \langle T_p, \mathcal{X}_p, \mathcal{N}_p \rangle$, where

1. $T_p \in [S]^{<\omega}$.

2. $\mathcal{X}_p \in [\mathcal{I}]^{<\omega}$.

3. $\mathcal{N}_p$ is a finite $\in$-chain of countable elementary submodels of some large enough model $(H_\theta, \in)$, containing the ideal $\mathcal{I}$.

$\mathcal{N}_p$ separates T_p is the following sense

4. $(\forall x, y \in T_P) \, (x \neq y) \, (\exists N \in \mathcal{N}_p) \, (|\{x, y\} \cap N| = 1)$.

5. $(\forall N \in \mathcal{N}_p) \, (\forall x \in T_p \smallsetminus N) \, (\forall Z \in \mathcal{P}(S) \cap N) \, (x \in Z \to [Z]^\omega \not\subseteq \mathcal{I})$.

We order $\mathcal{P}$ as follows: set $p \leq q$ iff

6. $T_p \supseteq T_q \; \mathcal{X}_p \supseteq \mathcal{X}_q, \; \mathcal{N}_p \supseteq \mathcal{N}_q$ and

7. $(\forall X \in \mathcal{X}_q) \, (T_p \smallsetminus T_q) \cap X = \emptyset)$.

In order to apply $\mathrm{mm} > \omega_1$, we need to show the following

Claim. $\mathcal{P}$ *is stationary-set preserving.*

Proof. Let $\langle C(p) : p \in \mathcal{P} \rangle$ be a $\mathcal{P}$-club. Let $\bar{p} \in \mathcal{P}$ and A a stationary subset of ω_1. We need to find $q \leq \bar{p}$ such that $C(q) \cap A \neq \emptyset$.

Let M be a countable elementary submodel of a large enough $(H_\lambda, \in)$, which contains $\langle C(p) : p \in \mathcal{P} \rangle$, $\bar{p}$ and A. We also require that $\delta = M \cap \omega_1 \in A$. Define $\bar{q} = \langle T_{\bar{p}}, \mathcal{X}_{\bar{p}}, \mathcal{N}_{\bar{p}} \cup \{M \cap H_\theta\} \rangle$. Then we have that $\bar{q} \leq \bar{p}$.

By $\mathcal{P}$-unboundedness, we can find $q \leq \bar{q}$ such that $C(q) \smallsetminus \delta \neq \emptyset$. Then we have the following

Subclaim. $\delta \in C(q)$.

Proof. Assume not. Then by $\mathcal{P}$-closedness, there is $\beta < \delta$ and $r \leq q$ such that

$$\forall s \leq r \quad C(s) \cap [\beta, \delta) = \emptyset \quad (*).$$

We will find $\bar{r} \in M \cap \mathcal{P}$ with the following properties

i) $\bar{r}$ end-extends $r \cap M$.

ii) $C(\bar{r}) \smallsetminus \beta \neq \emptyset$.

iii) $\bar{r}$ and r are compatible.

Then we will have that a common extension s of $\bar{r}$ and r does not satisfy $(*)$, so, by elementarity we get a contradiction. To define $\bar{r}$, let $H_r = T_r \smallsetminus (T_r \cap M)$ and $n = |H_r| \geq 1$. To construct the extension $\bar{r}$, we have to take care of H_r, so that condition (7) is satisfied.
   Define

$$\mathcal{H} = \{H \in [S]^n : (\exists s)(s \ end \ extending \ r \cap M)(H \subseteq T_s \smallsetminus T_{r \cap M})\}.$$

Observe that $\mathcal{H} \in M \cap H_\theta \in N$ for every $N \in \mathcal{N}_r \smallsetminus M$. Let $\mathcal{H}_0 = \mathcal{H}$, and for each $1 \leq i \leq n$ define

$$H_i = \{H \in [S]^{n-i} : \{x \in S : H \cup \{x\} \in \mathcal{H}_{i-1}\} \in \tilde{\mathcal{I}}^+\},$$

where $\tilde{\mathcal{I}} = \{X \subseteq S : [X]^\omega \subseteq \mathcal{I}\}$ and $\tilde{\mathcal{I}}^+ = \mathcal{P} \smallsetminus \tilde{\mathcal{I}}$.

Observe that each $\mathcal{H} \in M \cap H_\theta \in N$, for each $N \in \mathcal{N}_r$, and therefore $\mathcal{H}_i \in N$ for every $1 \leq i \leq n$ and $N \in \mathcal{N}_r$.
   Let $H_r = \{x_1^r, x_2^r, \ldots, x_n^r\}$ be an strictly increasing enumeration of H_r. We will prove the following:

Subclaim. $\emptyset \in \mathcal{H}_n$, $\{x_1^r\} \in \mathcal{H}_{n-1}$, $\{x_1^r, x_2^r, \ldots, x_n^r\} \in \mathcal{H}$.

Proof. We will prove it by induction. Suppose that $\{x_1^r, x_2^r, \ldots, x_{i+1}^r\} \in \mathcal{H}_{n-(i+1)}$. We will prove that $\{x_1^r, x_2^r, \ldots, x_i^r\} \in \mathcal{H}_{n-i}$.

Observe that

$$X = \{x \in S : \{x_1^r, x_2^r, \ldots, x_i^r, x\} \in \mathcal{H}_{n-(i+1)}\} \in N$$

for each $N \in \mathcal{N}_r$. Observe that by hypothesis $x_{i+1}^r \in X$. Let $N \in \mathcal{N}_r$ be a model such that $\{x_1^r, x_2^r, \ldots, x_i^r\} \in N$ and $x_{i+1}^r \notin N$. By condition (5), we have that $X \in \tilde{\mathcal{I}}^+$. So, by definition of $\mathcal{H}_{n-i}$, we have that $\{x_1^r, x_2^r, \ldots x_i^r\} \in \mathcal{H}_{n-i}$.

We in particular have that $\mathcal{H}_n = \{\emptyset\}$. $\square$

Define
$$X_1 = \{x : \{x\} \in \mathcal{H}_{n-1}\}$$

Then $X_1 \in M \cap H_\theta$. We need to find $x_1 \in X_1 \cap M \smallsetminus (\cup \mathcal{X}_r)$. If there is not such x_1, then we would have that

$$[X_1]^\omega \subseteq \mathcal{I}$$

i.e., $M \models [X_1]^\omega \subseteq \mathcal{I}$, but this contradicts the fact that $\emptyset \in \mathcal{H}_n$. Define now,
$$X_2 = \{x \in S : \{x_1, x\} \in \mathcal{H}_{n-2}\}.$$

Then $X_2 \in \tilde{\mathcal{I}}^+ \cap M$. By the previous argument we can choose $x_2 \in X_2 \smallsetminus \bigcup \mathcal{X}_r$, continue this procedure, we get $\{x_1, x_2, \ldots, x_n\} \in \mathcal{H}_\theta \cap M$ such that $x_i \notin \cup \mathcal{X}_r$ for every $1 \le i \le n$. Therefore we can define $\bar{r}$ end extending $r \cap M$ such that $T_{\bar{r}} \smallsetminus T_r \cap M = \{x_1, x_2, \ldots, x_n\}$.

Hence $\langle T_r \cup T_{\bar{r}}, \mathcal{X}_r \cup \mathcal{X}_{\bar{r}}, \mathcal{N}_r \cup \mathcal{N}_{\bar{r}} \rangle$ extends r and $\bar{r}$. As we mentioned above this is already a contradiction.

$\square$

This completes the proof of the claim. $\square$

Let us continue with the proof of Theorem 54. In order to get $T \subset S$ uncountable and orthogonal to $\mathcal{I}$ (so, the second condition holds), we define for every $X \in \mathcal{F}$ and $\alpha < \omega_1$ the following subsets of $\mathcal{P}$

. $\mathcal{D}_X = \{p \in \mathcal{P} : X \in \mathcal{X}_p\}$

. $\mathcal{D}_\alpha = \{p \in \mathcal{P} : (\exists N \in \mathcal{N}_p)(\alpha \in N, T_p \smallsetminus N \neq \emptyset)\}$

Notice that if $p \in \mathcal{P}$ is arbitrary, $q = \langle T_p, \mathcal{X}_p \cup \{X\}, \mathcal{N}_p \rangle \in \mathcal{D}_X$ and $q \le p$, so $\mathcal{D}_X$ is dense for each $X \in \mathcal{F}$.

To prove that $\mathcal{D}_\alpha$ is dense for each α. Fix $\alpha < \omega_1$ and $p \in \mathcal{P}$ ($p = \langle T_p, \mathcal{X}_p, \mathcal{N}_p \rangle$. Take N a countable elementary submodel of (H_θ, ϵ) containing α and p. Let $I = \bigcup \mathcal{X}_p$. Since the ideal is proper we have that $S \smallsetminus I \neq \emptyset$. Furthermore $S \smallsetminus I$ cannot be covered by countable many sets S_n such that $[S_n]^\omega \subseteq \mathcal{I}$, then we can find $x \in S \smallsetminus I \smallsetminus N$, such that $x \notin Z$ for every $Z \in N \cap \mathcal{P}(S)$ such that $[Z]^\omega \subseteq \mathcal{I}$. Therefore $\langle T_p \cup \{x\}, \mathcal{X}_p, \mathcal{N}_p \cup \{N\}\rangle$ is an extension of p in $\mathcal{D}_\alpha$. Now, applying $\mathfrak{mm} > \omega_1$, we can find a generic filter $\mathcal{G} \subseteq \mathcal{P}$, intersecting all the dense sets listed above. Finally, define $T = \bigcup_{p \in \mathcal{G}} T_p$. If we suppose that $T = \{s_m : m \in \mathbb{N}\}$ (i.e. T is countable), then we can find a countable collection N_m, such that $T \subseteq \bigcup_{m=1}^{\infty} N_m$, where $s_n \in T_m \subseteq N_m$ and $p_m = \langle T_m, \mathcal{X}_m, \mathcal{N}_m \rangle \in \mathcal{G}$, and $s_m \in T_m \subseteq N_m \in \mathcal{N}_m$. But then, there should be α not in $\bigcup_{m=1}^{\infty} N_m$. Since $\mathcal{G} \cap \mathcal{D}_\alpha \neq \emptyset$, we can find $p \in \mathcal{G}$ and $N_p \in \mathcal{N}_p$, such that $\alpha \in N_p$ and $T_p \smallsetminus N_p \neq \emptyset$, which is a contradiction, since p and p_m are elements in $\mathcal{G}$ which do not have a common extension. Hence we have that T is uncountable. To prove that T is orthogonal to $\mathcal{F}$, take $F \in \mathcal{F}$ and $p \in \mathcal{G} \cap \mathcal{P}$, such that $I \in \mathcal{X}_P$. We prove that $T \cap F \subseteq T_p$, hence it is finite. Take $s \in T \cap I$, then there is $q \in \mathcal{G}$ such that $s \in T_q$. Let $r \in G$ such that $r \leq p, q$. Then we have that $s \in T_r$. Since $r \leq p$, and since $I \in \mathcal{X}_p$, we have that $(T_r \smallsetminus T_p) \cap I = \emptyset$, then $s \in T_p$. So, we have that condition 1 of the dichotomy holds.

$\square$

Corollary 55. $\mathfrak{mm} > \omega_1$ *implies* **SID**.

8.2 Sparse set-mapping principle

Definition. For a given set S, a *set mapping* is a map $F : S \to \mathcal{P}(S)$. We say that $T \subseteq S$ is *F-free* if $F(x) \cap T \subseteq \{x\}$ for every $x \in T$.

Definition. A set mapping $F : S \to \mathcal{P}(S)$ is *sparse* if there is a proper σ-ideal $\mathcal{I}$ on S such that for every $Z \in \mathcal{I}^+$ there exists $Z_0 \in [Z]^{\leq \omega}$ and $H \in \mathcal{I}$ such that $Z_0 \not\subseteq \bigcup_{x \in a} F(x)$ for all $a \subseteq S \smallsetminus H$ finite.

We will denote by **SMP** the following statement:

(**SMP**) Every sparse set-mapping $F : S \to \mathcal{P}(S)$ admits an uncountable F-free set $T \subseteq S$.

We will prove that **SMP** is a consequence of $\mathfrak{mm} > \omega_1$.

Theorem 56. *If* $\mathfrak{mm} > \omega_1$ *then* **SMP** *holds*.

Proof. Let $\mathcal{I}$ a fixed σ-ideal witnessing the sparseness of $F : S \to \mathcal{P}(S)$. We will construct an uncountable F-free set, $T \subset S$. Let $\mathcal{P}$ be the collection of all pairs of the form $\langle T_p, \mathcal{N}_p \rangle$, where

1. $T_p \in [S]^{<\omega}$.

2. $\mathcal{N}_p$ is a finite ϵ-chain of countable elementary submodels of a fixed large enough H_θ.

3. $(\forall x, y \in T_p)\, (x \neq y)\, (\exists N \in \mathcal{N}_p)(|\{x, y\} \cap N| = 1)$.

4. $(\forall N \in \mathcal{N}_p)\, (\forall x \in T_p \smallsetminus N)\, (\forall X \in \mathcal{I} \cap N)\, (x \notin X)$.

order $\mathcal{P}$ as follows: set $p \leq q$ iff

5. $T_p \supseteq T_q$, $\mathcal{N}_p \supseteq \mathcal{N}_q$ and

6. $\forall x \in T_q\, F(x) \cap (T_p \smallsetminus T_q) = \emptyset$

To apply $\mathfrak{mm} > \omega_1$, we will prove the following

Claim. *$\mathcal{P}$ is stationary-set preserving.*

Proof. The structure of the proof is essentially the same as the proof of the first claim in the previous theorem. Let $\langle C(p) : p \in \mathcal{P} \rangle$ be a $\mathcal{P}$-club. Let $\bar{p} \in \mathcal{P}$ and A a stationary subset of ω_1. We need to find $q \leq \bar{p}$ such that $C(q) \cap A \neq \emptyset$.

Let M be a countable elementary submodel of a large enough (H_λ, ϵ), which contains $\langle C(p) : p \in \mathcal{P} \rangle$, $\bar{p}$ and A, and M such that $\delta = \omega_1 \cap M \in A$. Define $\bar{q} = \langle T_{\bar{p}}, \mathcal{N}_{\bar{p}} \cup \{M \cap H_\theta\} \rangle$. Then we have that $\bar{q} \leq \bar{p}$.

By $\mathcal{P}$-unboundedness, we can find $q \leq \bar{q}$ such that $C(q) \smallsetminus \delta \neq \emptyset$. Then we have the following

Subclaim. $\delta \in C(q)$.

Proof. Assume not. Then by $\mathcal{P}$-closedness, there is $\beta < \delta$ and $r \leq q$ such that

$$C(s) \cap [\beta, \delta) = \emptyset \text{ for every } s \leq r \text{ (*)}.$$

We will find $\bar{r} \in M \cap \mathcal{P}$ with the following properties

i) $\bar{r}$ end-extends $r \cap M$.

ii) $C(\bar{r}) \smallsetminus \beta \neq \emptyset$.

iii) $\bar{r}$ and r are compatible.

Then we will have that a common extension s of $\bar{r}$ and r does not satisfy (*), getting a contradiction. To define $\bar{r}$, let $H_r = T_r \smallsetminus (T_r \cap M)$ and $n = |H_r| \geq 1$.

Define

$$\mathcal{H} = \{H \in [S]^n : (\exists s)(s \ \text{end extending} \ r \cap M)(H \subseteq T_s \smallsetminus T_{r \cap M})\}.$$

Observe that $\mathcal{H} \in M \cap H_\theta \in N$ for every $N \in \mathcal{N}_r \smallsetminus M$.

Let $\mathcal{H}_0 = \mathcal{H}$, and for each $1 \leq i \leq n$ define

$$H_i = \{H \in [S]^{n-i} : \{x \in S : H \cup \{x\} \in \mathcal{H}_{i-1}\} \in \mathcal{I}^+\},$$

where $\mathcal{I}^+ = \mathcal{P} \smallsetminus \mathcal{I}$.

Observe that each $\mathcal{H} \in M \cap H_\theta \in N$, for each $N \in \mathcal{N}_r$, and therefore $\mathcal{H}_i \in N$ for every $1 \leq i \leq n$ and $N \in \mathcal{N}_r$.

Let $H_r = \{x_1^r, x_2^r, \ldots, x_n^r\}$ be a strictly increasing enumeration of H_r. We need to prove the following:

Subclaim. $\emptyset \in \mathcal{H}_n, \{x_1^r\} \in \mathcal{H}_{n-1}, \{x_1^r, x_2^r, \ldots, x_n^r\} \in \mathcal{H}$.

Proof. We will prove it by induction. Suppose that $\{x_1^r, x_2^r, \ldots, x_{i+1}^r\} \in \mathcal{H}_{n-(i+1)}$. We will prove that $\{x_1^r, x_2^r, \ldots, x_i^r\} \in \mathcal{H}_{n-i}$.

Observe that

$$X = \{x \in S : \{x_1^r, x_2^r, \ldots, x_i^r, x\} \in \mathcal{H}_{n-(i+1)}\} \in N$$

for each $N \in \mathcal{N}_r$. Let $N \in \mathcal{N}_r$ be a model such that $\{x_1^r, x_2^r, \ldots x_i^r\} \in N$ and $x_{i+1}^r \notin N$. By hypothesis, we have that $x_{i+1}^r \in X$. Hence, by condition (4), we must have that $X \in \tilde{\mathcal{I}}^+$. So, by definition of $\mathcal{H}_{n-i}$, we have that $\{x_1^r, x_2^r, \ldots x_i^r\} \in \mathcal{H}_{n-i}$ (Fubini's Argument).

We in particular have that $\mathcal{H}_n = \{\emptyset\}$. $\qquad\qquad\square$

Define
$$X_1 = \{x : \{x\} \in \mathcal{H}_{n-1}\}.$$

Then $X_1 \in M \cap H_\theta$. We need to find $x_1 \in X_1 \cap M \smallsetminus F''T_r$. If there is not such x_1, then we would have that

$$X_1 \in \mathcal{I}$$

i.e., $M \models X_1 \in \mathcal{I}$, but this contradicts the fact that $\emptyset \in \mathcal{H}_n$. Define now,

$$X_2 = \{x \in S : \{x_1, x\} \in \mathcal{H}_{n-2}\}.$$

Then $X_2 \in \mathcal{I}^+ \cap M$. By the previous argument we can choose $x_2 \in X_2 \smallsetminus F''T_r$, continue this procedure, we get $\{x_1, x_2, \ldots, x_n\} \in \mathcal{H}_\theta \cap M$

such that $x_i \notin F''T_r$ for every $1 \leq i \leq n$. Therefore we can define $\bar{r}$ end extending $r \cap M$ such that $T_{\bar{r}} \smallsetminus T_r \cap M = \{x_1, x_2, \ldots, x_n\}$.

Hence $\langle T_r \cup T_{\bar{r}}, \mathcal{N}_r \cup \mathcal{N}_{\bar{r}} \rangle$ extends r and $\bar{r}$.

□

This completes the proof of the claim.   □

We define for every $\alpha < \omega_1$ the following subsets of $\mathcal{P}$

. $\mathcal{D}_\alpha = \{p \in \mathcal{P} : (\exists N \in \mathcal{N}_p)(\alpha \in N, T_p \smallsetminus N \neq \emptyset)\}$

Now, if we define $T = \bigcup_{p \in \mathcal{G}} T_p$, we have that T is an uncountable F-free set.   □

8.3 *P*-ideal dichotomy

Definition. We say that an ideal $\mathcal{I}$ on a given set S is a *P-ideal* if for every family $\{I_n : n \in \mathbb{N}\} \subseteq I$ there is $I \in \mathcal{I}$ such that $I_n \subseteq^* I$ for every $n \in \mathbb{N}$.

This leads us to the following dual form of the Small Ideal Dichotomy, SID, considered above.

Definition. (PID) Let $\mathcal{I}$ a *P*-ideal of countable subsets of some set S, then either

1. There is an uncountable $T \subset S$ such that $[T]^{\leq \omega} \subseteq \mathcal{I}$, or

2. There is a decomposition $S = \bigcup_{n=1}^{\infty} S_n$ such that S_n is orthogonal to $\mathcal{I}$ for every $n \in \mathbb{N}$.

Theorem 57. $\mathrm{mm} > \omega_1$ *implies* **PID**.

Proof. Fix a *P*-ideal $\mathcal{I}$ of countable subsets of S and assume that the second alternative of fails, or in other words, S cannot be decomposed into countably many sets orthogonal to $\mathcal{I}$. We shall find a stationary-set preserving (in fact, proper) poset $\mathcal{P}$ which forces an uncountable set $X \subseteq S$ such that $[X]^{\aleph_0} \subseteq \mathcal{I}$.

Fix a sufficiently large regular cardinal θ such that the structure $(H_\theta, \in)$ contains the set $[S]^{\aleph_0}$. For a countable elementary submodel $N \prec H_\theta$, we fix $b_N \in \mathcal{I}$ such that $b_N \subseteq S \cap N$ and

$$(\forall a \in \mathcal{I} \cap N) \ a \subseteq^* b_N,$$

and we fix a point $x_N \in S$ such that

$$(\forall X \in \mathcal{P}(S) \cap N)[X \perp \mathcal{I} \Rightarrow x_N \notin X].$$

Expanding $(H_\theta, \in)$ by adding a well-ordering $<_\theta$ of H_θ we may assume $N \mapsto b_N$ and $N \mapsto x_N$ are definable maps.

Let $\mathcal{P}$ be the collection of all finite $\in$-chains p of countable elementary submodels of $(H_\theta, \in, <_\theta)$. For $p, q \in \mathcal{P}$, set $p \leq q$ if and only if,

(a) $p \supseteq q$, and

(b) $(\forall M \in q)(\forall N \in M \cap (p \setminus q))[x_N \in b_M]$.

Claim 1. $\mathcal{P}$ is a stationary-set preserving poset.

Proof. Let $(C(p) : p \in \mathcal{P})$ be a given $\mathcal{P}$-club, $\bar{p}$ a given condition of $\mathcal{P}$ ane E a given stationary set. We need to find an extension q of $\bar{p}$ such that $C(q) \cap E \neq \emptyset$.

Let $\kappa = (2^\theta)^+$ and choose a countable elementary submodel $M \prec (H_\kappa, \in)$ containing

$$(H_\theta, \in <_\theta), \mathcal{P}, S, \mathcal{I}, \bar{p}$$

such that $\delta = M \cap \omega_1 \in E$. Let

$$p = \bar{p} \cup \{M \cap H_\theta\}.$$

Choose $q \leq p$ such that $C(q) \setminus \delta \neq \emptyset$. We claim that $\delta \in C(q)$ which will finish the proof. Otherwise, there is $r \leq q$ and $\gamma < \delta$ such that

$$(\forall s \leq r) C(s) \cap (\gamma, \delta] = \emptyset.$$

We may assume $\gamma > N \cap \omega_1$ for any model of $r \cap M$. Let $\mathcal{X}$ be the collection of all $x \in \mathcal{P}$ that are realizing the same time over the parameters γ and $r \cap M$, and so in particular, end-extend $r \cap M$. As before to get the desired contradiction it suffices to find $\bar{r} \in \mathcal{X}$ compatible with r.

For $s \in \mathcal{D}$ end-extending $r \cap M$, let

$$x^s = \langle x_1^s, x_2^s, \ldots, x_l^s \rangle$$

where

$$\{x_N : N \in s \setminus (r \cap M)\} = \{x_1^s, x_2^s, \ldots, x_l^s\}$$

enumerated according to the $\in$-ordering of s. Let k be the length of the sequence x^r associated to the condition r which clearly end-extends $r \cap M$. Let

$$\mathcal{F} = \left\{ x \in S^k : (\exists s \sqsupseteq r \cap M)\, [s \in \mathcal{X} \ \& \ x_s = x] \right\}.$$

Then $\mathcal{F} \in M$ and $x^r = \langle x_{N_1}, x_{N_2}, \dots, x_{N_k} \rangle \in \mathcal{F}$. It follows that $\mathcal{F}$ is *large* relative to the co-ideal

$$\mathcal{H} = \{ X \subseteq S : \mathcal{I} \restriction X \not\models \text{ alternative } (2) \}.$$

So in particular $\mathcal{F}$ contains a subfamily $\mathcal{F}_0 \in M$ that form the set of maximal nodes of an everywhere $\mathcal{H}$-branching subtree of $S^{\leq k}$ of height k. Let

$$X_1 = \{ \xi \in S : (\exists x \in \mathcal{F}_0)\, \xi = x_1 \}.$$

Then $X_1 \in M \cap \mathcal{H}$. So there is infinite $a_1 \in \mathcal{I} \cap M$ such that $a_1 \subseteq X_1$. Then $a_1 \subseteq^* b_{N_i}$ for all $i = 1, 2, \dots, k$. So we can pick $\xi_1 \in a_1$ such that $\xi_1 \in \bigcap_{i=1}^{k} b_{N_i}$. Let

$$X_2 = \{ \xi \in S : (\exists x \in \mathcal{F}_0)\, [\xi_1 = x_1 \ \& \ \xi = x_2] \}.$$

Then $X_2 \in M \cap \mathcal{H}$. So there must be infinite $a_2 \in \mathcal{I} \cap M$ such that $a_2 \subseteq X_2$. Then again $a_2 \subseteq^* \bigcap_{i=1}^{k} b_{N_i}$, so we can choose $\xi_2 \in a_2 \cap \bigcap_{i=1}^{k} b_{N_i}$, and so on. At the end of this process one obtains $\{ \xi_1, \xi_2, \dots, \xi_k \} \subseteq \bigcap_{i=1}^{k} b_{N_i}$ such that

$$\langle \xi_1, \xi_2, \dots, \xi_k \rangle \in \mathcal{F} \cap M.$$

Pick $\bar{r} \in \mathcal{D} \cap M$ end-extending $r \cap M$ such that $x_{\bar{r}} = \langle \xi_1, \xi_2, \dots, \xi_k \rangle$. Then r and $\bar{r}$ are two compatible conditions of $\mathcal{P}$ and, in fact, $r \cup \bar{r}$ is their common extension. □

□

Exercise. Show that **PID** implies $\mathfrak{b} \leq \aleph_2$.

Problem 8.3.1. Does **PID** imply that $\mathfrak{c} \leq \aleph_2$.

Exercise. Show that **PID** implies $\theta^{\aleph_0} = \theta$ for all regular $\theta \geq 2^{\aleph_0}$

Exercise. Show that **PID** implies **SID** for ideals generated by $< \mathfrak{p}$ sets.

Exercise. Let $\mathcal{J}$ be an ideal of countable subsets of some set S that is generated by $< \mathfrak{p}$ sets. Let $\mathcal{I}$ be its *orthogonal*, i.e., the ideal of countable subsets of S that have finite intersections with all sets from $\mathcal{J}$. Show that $\mathcal{I}$ is a P-ideal and that **SID**$(\mathcal{J})$ implies **PID**$(\mathcal{I})$.

Chapter 9

Coherent and Lipschitz Trees

9.1 The Lipschitz condition

Given a tree T, let $\Delta : T^2 \to \mathrm{Ord}$ be defined by

$$\Delta(s,t) = \mathrm{otp}\{x \in T : x <_T s \text{ and } x <_T t\}.$$

One should view Δ as some sort of distance function on T by interpreting inequalities like $\Delta(x,y) > \Delta(x,z)$ as saying that x is closer to y than to z. A partial map g from a tree S into a tree T is *Lipschitz*, if g is level-preserving and

$$\Delta(g(x), g(y)) \geq \Delta(x,y)$$

for all $x, y \in \mathrm{dom}(g)$.

Definition. A *Lipschitz tree* is any Aronszajn tree T with the property that every level-preserving map from an uncountable subset of T into T is Lipschitz on an uncountable subset of its domain.

Definition. A *coherent tree* is any Aronszajn tree isomorphic to a downwards closed subset T of $I^{<\omega_1}$, for some countable index set I such that for all $s, t \in T$ of the same height α, the set

$$\{\xi < \alpha : s(\xi) \neq t(\xi)\}$$

is finite.

The following fact shows that a typical coherent tree is satisfies the Lipschitz condition and that therefore Lipschitz trees appear in profusion.

Theorem 58. *A coherent tree T is Lipschitz if and only if every uncountable subset of T contains an uncountable antichain.*

Proof. The converse implication is easy and is therefore left to the reader, so let us concentrate on proving the direct implication. Let T be a given coherent downwards closed subtree of the tree of all maps from countable ordinals into ω with the property that every uncountable subset of T contains an uncountable antichain. Thus, the height of a given node t of T is equal to its domain. Similarly, for two different nodes $s, t \in T$ of the same height α,

$$\Delta(s,t) = \mathrm{otp}\{x \in T : x <_T s \text{ and } x <_T t\} = \min\{\xi < \alpha : s(\xi) \neq t(\xi)\}.$$

Consider a partial level-preserving map f from an uncountable subset X of T into T. By our assumption, we may assume that both domain and the range of f are antichains of T. Moreover, we may assume that X contains no two different nodes of the same height. For $t \in X$, set

$$D_t = \{\xi < \mathrm{ht}(t) : t(\xi) \neq f(t)(\xi)\}.$$

Applying the Δ-system lemma and shrinking X, if necessary, we may assume that D_t ($t \in X$) forma a Δ-system with root D. Shrinking X even further, we may assume that:

(1) $D_s \setminus D < D_t \setminus D$ whenever $s, t \in X$ are such that $\mathrm{ht}(s) < \mathrm{ht}(t)$,

(2) $s \restriction (\max(D) + 1) = t \restriction (\max(D) + 1)$ for all $s, t \in X$.

Note that if $D_t = D$ for all $t \in X$, then (1) and (2) ensure that

$$\Delta(s,t) = \Delta(f(s), f(t)) \text{ for all } s \neq t \text{ in } X, \tag{9.1}$$

and so, in particular, the map f is Lipschitz. Thus we may assume that $D_t \setminus D \neq \emptyset$ for all $t \in X$. For $t \in X$, let $\delta_t = \min(D_t \setminus D)$. By our assumption about T, we can find an uncountable subset Y of X such that both sets

$$\{t \restriction \delta_t : t \in Y\} \text{ and } \{f(t) \restriction \delta_t : t \in Y\}$$

are antichains of T. It follows again that $\Delta(s,t) = \Delta(f(s), f(t))$ for all $s \neq t$ in X, as required. $\square$

The following property of Lipschitz trees will be quite frequently and implicitly used in this and the following section.

Theorem 59. *Suppose T is a Lipschitz tree, n is a positive integer, and that A is an uncountable family of pairwise disjoint n-element subsets of T. Then there exists an uncountable $B \subseteq A$ such that $\Delta(a_i, b_i) = \Delta(a_j, b_j)$ for all $a \neq b$ in B and $i, j < n$.*

Proof. Replacing the family A by a family of the form $a \restriction \xi_a$ $(a \in A)$ and applying Theorem 58, we may assume that a given member of A is in fact a subset of some level of T. Fix $i, j < n$ and apply the assumption that T is a Lipschitz tree to the partial level-preserving map

$$a_i \longmapsto a_j \ (a \in A)$$

obtaining an uncountable $A_0 \subseteq A$ such that

$$\Delta(a_i, b_i) \leq \Delta(a_j, b_j) \text{ for all } a, b \in A_0.$$

Applying the assumption that T is a Lipschitz tree to the inverse map $a_j \longmapsto a_i$ $(a \in A_0)$ will give us an uncountable $A_1 \subseteq A$ such that

$$\Delta(a_i, b_i) = \Delta(a_j, b_j) \text{ for all } a, b \in A_1.$$

Repeating this procedure successively for every pair $i, j < n$, we reach the conclusion of Theorem 59. $\qquad\qquad\square$

The following result shows that under some mild axiomatic assumption the class of Lipschitz trees in fact coincides with the class of coherent trees, the main object of our study here.

Theorem 60. *Under* $\mathfrak{m} > \omega_1$, *every Lipschitz tree T is isomorphic to a coherent downwards closed subtree of the tree of all maps from countable ordinals into* ω.

Proof. Define $\mathcal{P}$ to be the poset of all finite partial functions p from $T \times \omega_1$ into ω such that the following holds for all $x \neq y$ in $\mathrm{dom}_0(p)$[1]:

(1) $p(x, \xi) = p(y, \xi)$ for $\xi < \Delta(x, y)$,

(2) $p(x, \xi) \neq p(y, \xi)$ for $\xi = \Delta(x, y)$[2].

We let p extend q, if p extends q as a function, and

(3) $p(x, \xi) = p(y, \xi)$ for all $x, y \in \mathrm{dom}_0(q)$ and $\xi < \mathrm{ht}(x), \mathrm{ht}(y)$ with the property that $\xi \notin \mathrm{dom}_1(q)$.

It is clear that a sufficiently generic filter will give us the desired embedding, and so we concentrate on showing that $\mathcal{P}$ satisfies the countable chain condition.

---

[1]Here and below, $\mathrm{dom}_0(p) = \{t \in T : (t, \alpha) \in \mathrm{dom}(p) \text{ for some } \alpha\}$ and $\mathrm{dom}_1(p) = \{\xi \in \omega_1 : (t, \xi) \in \mathrm{dom}(p) \text{ for some } t \in T\}$.

[2]In (1) and (2) we are making the implicit requirement that for every $\xi \in \mathrm{dom}_1(p)$ and $x \neq y \in \mathrm{dom}_0(p)$, if $\xi \leq \Delta(x, y)$, then $(x, \xi) \in \mathrm{dom}(p)$ iff $(y, \xi) \in \mathrm{dom}(p)$, and moreover, that always $(x, \Delta(x, y))$ and $(y, \Delta(x, y))$ belong to $\mathrm{dom}(p)$.

So let p_δ $(\delta < \omega_1)$ be a given sequence of elements of $\mathcal{P}$. Let a_δ be the projection of $\mathrm{dom}_0(p_\delta)$ onto the δth level of T and let

$$h(\delta) = \max\{\Delta(s,t) + 1 : s, t \in a_\delta, s \neq t\}.$$

Then there is a stationary set Γ of countable limit ordinals on which the mapping h, as well as the mapping $\delta \longmapsto \mathrm{dom}(p_\delta) \cap (T \upharpoonright \delta) \times \delta$ is constant. Let α and F be these constant values respectively. We may assume that all a_δ's project to the same set a on the αth level and all p_δ's generate isomorphic structures over α, a and F. Thus, in particular we want that the isomorphism between the p_δ's $(\delta \in \Gamma)$ respects a fixed enumeration $a_\delta(i)$ $(i < n)$ of a_δ, where n is the common cardinality of these sets. As in the previous proof, we find an uncountable subset Σ of Γ such that for all $\gamma \neq \delta$ in Σ:

(4) $a_\gamma(i)$ and $a_\delta(j)$ are incomparable for all $i, j < n$,

(5) $\Delta(a_\gamma(i), a_\delta(i)) = \Delta(a_\gamma(j), a_\delta(j))$ for all $i, j < n$.

We claim that if $\gamma \neq \delta$ are in Σ then p_γ and p_δ are compatible in $\mathcal{P}$. By (5), we have an ordinal β smaller than both γ and δ such that

$$\Delta(a_\gamma(i), a_\delta(i)) = \beta \text{ for all } i < n.$$

Define $r \in \mathcal{P}$ by letting its domain be

$$\mathrm{dom}(p_\gamma) \cup \mathrm{dom}(p_\delta) \cup \{(t, \beta) : t \in \mathrm{dom}_0(p_\gamma) \cup \mathrm{dom}_0(p_\delta), \mathrm{ht}(t) > \beta\}$$

and letting $r(t, \beta) = 0$ if $t \in \mathrm{dom}_0(p_\gamma)$, $\mathrm{ht}(t) > \beta$ and $r(t, \beta) = 1$ if $t \in \mathrm{dom}_0(p_\delta)$, $\mathrm{ht}(t) > \beta$. Note that r is indeed a member of $\mathcal{P}$, as it clearly satisfies the conditions (1) and (2) above. It is also easily checked that r extends both p and q, i.e. $r(x, \xi) = r(y, \xi)$ for all $x, y \in \mathrm{dom}_0(p)$ or $x, y \in \mathrm{dom}_0(q)$ and ξ is equal to β, the only new member of $\mathrm{dom}_1(r)$. This finishes the proof. $\qquad\square$

9.2   Filters and trees

Definition. For a given tree T of height ω_1,

$$\mathcal{U}(T) = \{A \subseteq \omega_1 : A \supseteq \Delta(X) \text{ for some uncountable } X \subseteq T\}.$$

The following fact isolates an important property of the class of Lipschitz trees and it will play the crucial role in the rest of this Chapter

Theorem 61. *The family $\mathcal{U}(T)$ is a uniform filter on ω_1 for every Lipschitz tree T. In particular, $\mathcal{U}(T)$ is a uniform filter on ω_1 for every coherent tree T with the property that every uncountable subset of T has an uncountable antichain.*

Proof. Given two uncountable subsets X and Y of T, we need to find an uncountable subset Z of T such that

$$\Delta(X) \cap \Delta(Y) \supseteq \Delta(Z).$$

By the assumption about T, it is clear that we may replace X and Y by two level-sequences x_δ ($\delta \in \Gamma$) and y_δ ($\delta \in \Gamma$) indexed by the same uncountable set Γ of limit ordinals. Moreover, we may assume that the x_δ's and y_δ's are all pairwise incomparable (see Theorem 58). Apply Theorem 59 to the subset (x_δ, y_δ) ($\delta \in \Gamma$) of $T \otimes T$ and obtain an uncountable set $\Sigma \subseteq \Gamma$ such that

$$\Delta(x_\gamma, x_\delta) = \Delta(y_\gamma, y_\delta) \text{ for all } \gamma, \delta \in \Sigma, \gamma \neq \delta.$$

So we can take Z to be any of the sets $\{x_\delta : \delta \in \Sigma\}$ or $\{y_\delta : \delta \in \Sigma\}$. This finishes the proof. $\qquad\square$

Theorem 62. *Assuming* $\mathfrak{m} > \omega_1$, *the filter* $\mathcal{U}(T)$ *is an ultrafilter for every coherent tree* T *with no uncountable branches.*

Proof. Let Γ be a given subset of ω_1. We need to find an uncountable subset X of T such that $\Delta(X)$ is included in either Γ or its complement. Let $\mathcal{P}_\Gamma$ be the poset of all finite subsets p of T that take at most one point from a given level of T such that

$$\Delta(p) = \{\Delta(x, y) : x, y \in p, x \neq y\} \subseteq \Gamma.$$

If $\mathcal{P}_\Gamma$ satisfies the countable chain condition, then a straightforward application of (*) gives an uncountable $X \subseteq T$ such that $\Delta(X) \subseteq \Gamma$. So let us consider the alternative that $\mathcal{P}_\Gamma$ fails to satisfy this condition. Let p_δ ($\delta < \omega_1$) be a sequence of pairwise incomparable members of $\mathcal{P}_\Gamma$. Re-enumerating, we may assume that every node of a given p_δ has height at least δ, so we can define $a_\delta =$ the projection of p_δ on the δth level of T. For $\delta \in \Lambda$, let

$$h(\delta) = \{\Delta(x, y) + 1 : x, y \in a_\delta, x \neq y\}.$$

Find stationary $\Omega \subseteq \Lambda$ such that h is constant on Ω. Let $\bar{\xi}$ be the constant value of h. Going to an uncountable subset of Ω, we may assume that all a_δ ($\delta \in \Omega$) are of equal size n and that they are given with an enumeration $a_\delta(i)$ ($i < n$). Moreover, we may assume that $a_\gamma(i) \upharpoonright \bar{\xi} = a_\delta(i) \upharpoonright \bar{\xi}$ for all $\gamma, \delta \in \Omega$. Applying Theorem 59, we obtain an uncountable $\Sigma \subseteq \Omega$ such that for all $\gamma \neq \delta$ in Σ:

(1) $a_\gamma(i)$ and $a_\delta(j)$ are incomparable for all $i, j < n$,

(2) $\Delta(a_\gamma(i), a_\delta(i)) = \Delta(a_\gamma(j), a_\delta(j))$ for all $i, j < n$.

It follows that for all $\gamma \neq \delta$ in Σ:

(3) $\Delta(p_\gamma \cup p_\delta) = \Delta(p_\gamma) \cup \Delta(p_\delta) \cup \{\Delta(a_\gamma(0), a_\delta(0))\}$.

Since $\Delta(p_\gamma)$ and $\Delta(p_\delta)$ are subsets of Γ, $\Delta(a_\gamma(0), a_\delta(0)) \notin \Gamma$ must hold. This gives rise to an uncountable set $X = \{a_\delta(0) : \delta \in \Sigma\}$ with the property that $\Delta(X) \cap \Gamma = \emptyset$. This finishes the proof. $\hspace{2em}\square$

The definition of $\mathcal{U}(T)$ can be relativized to any countable elementary submodel M of $(H_{\omega_2}, \in)$ as follows.

Definition. For a coherent tree T of height ω_1 and a countable elementary submodel M of H_{ω_2} with $T \in M$, let $\mathcal{U}_M(T)$ be the collection of all subsets A of ω_1 for which one can find uncountable $X \subseteq T$ with $X \in M$ and $t \in T_{M \cap \omega_1}$ belonging to the downward closure of X such that $A \supseteq \Delta(t, X) = \{\Delta(t, x) : x \in X\}$.

It should be clear that the basic idea of the proof of Theorem 61 also gives the following.

Theorem 63. *For every coherent tree T and every countable elementary submodel M of H_{ω_2} such that $T \in M$, the family $\mathcal{U}_M(T)$ is a nonprincipal filter concentrating on the set $M \cap \omega_1$.* $\hspace{2em}\square$

9.3   Model rejecting a finite set of nodes

From now on, we fix a special coherent subtree T of the complete binary tree $2^{<\omega_1}$ that is closed under restrictions and finite changes of its elements. We also fix a subset K of T and use K^c to denote its complement in T. Recall that for $x, y \in T$ by $x \wedge y$ one denotes the maximal common predecessor of x and y, and for a subset X of T one puts

$$\wedge(X) = \{x \wedge y : x, y \in X, x \neq y\}.$$

Our goal is to find an uncountable subset X of T such that either $\wedge(X) \subseteq K$ and $\wedge(X) \subseteq K^c$. Note that in Theorem 62, assuming $\mathfrak{m} > \omega_1$, we have achieved this in case when K is the union of levels of T. The general case requires a further analysis and a stronger axiom. For $J \in \{K, K^c\}$ and $t \in T$, the corresponding fiber is denoted by,

$$J(t) = \{\xi < \omega_1 : t \restriction \xi \in J\}.$$

For a subset X of T and $J \in \{K, K^c\}$, let $J(X) = \bigcup_{t \in X} J(t)$. The filters of the form $\mathcal{U}_M(T)$ are used to define the following important notion.

Definition. A countable elementary submodel M of H_{ω_2} *rejects* a finite subset F of T if $K^c(F) \in \mathcal{U}_M(T)$.

Note that if an elementary submodel M of H_{ω_2} such that $\{T, K\} \subseteq M$ rejects a singleton $\{t\}$ then there is an uncountable antichain X of T such that $\wedge(X) \subseteq K^c$. Note also that if M and N are two countable elementary submodels containing T and K such that $N \subseteq M$ and $M \cap \omega_1 = N \cap \omega_1$, and if N rejects F, then so does M. This follows from the inclusion $\mathcal{U}_N(T) \subseteq \mathcal{U}_M(T)$.

The following lemma suggests the direction of further analysis as it is definitely saying something about the poset of all finite approximations to an uncountable subset X of T such that $\wedge(X) \subseteq K$.

Theorem 64. *Suppose we have an uncountable family A of pairwise disjoint finite subsets of T all of some fixed size n such that the collection $\mathcal{M}$ of all countable elementary submodels of H_{ω_2} which do not reject any member of the family A is a stationary subset of $[H_{\omega_2}]^\omega$. Then there exist $a \neq b$ in A such that $a(i) \wedge b(i) \in K$ for all $i < n$.*

Proof. Using Theorem 58 and going to an uncountable subfamily of A, we may assume that

$$\Delta(a_i, b_i) = \Delta(a_j, b_j) \text{ for all } a \neq b \text{ in } A \text{ and } i, j < n. \qquad (9.2)$$

Since $\mathcal{M}$ is stationary, we can find $M \in \mathcal{M}$ such that $\{A, T, K\} \subseteq M$. Pick a $b \in A$ such that the height of every node from b is bigger that $\delta = M \cap \omega_1$. Let $X = \{a(0) : a \in A\}$ and let $t = b(0) \upharpoonright \delta$. Then $\Delta(t, X)$ is a typical generator of $\mathcal{U}_M(T)$ and by our assumption that M does not reject b, we conclude that

$$\Delta(t, X) \cap \delta \nsubseteq K^c(b). \qquad (9.3)$$

Applying (9.2) and (9.3), we conclude that there must be $a \in A \cap M$ such that $a(i) \wedge b(i) \in K$ for all $i < n$. $\qquad \square$

We shall need the following corollary of this Lemma.

Theorem 65. *Assuming $\mathfrak{m} > \omega_1$, if there is an uncountable subset Y of T such that the collection $\mathcal{M}$ of all countable elementary submodels of H_{ω_2} which do not reject any finite subset of Y is stationary in $[H_{\omega_2}]^\omega$, then there is uncountable $X \subseteq Y$ such that $\wedge(X) \subseteq K$.*

Proof. By Theorem 64, the poset of all finite $F \subseteq Y$ such that $\wedge(F) \subseteq K$ satisfies the countable chain condition. $\qquad \square$

Let $\theta_0 = \aleph_2(= (\aleph_1)^+)$ and let $\mathcal{C}_0$ be the collection of all closed unbounded subsets of $[H_{\theta_0}]^\omega$ consisting of countable elementary submodels M of $(H_{\theta_0}, \in)$ such that $\{T, K\} \subseteq M$. Similarly, let $\theta_1 = (2^{\aleph_1})^+$ and let $\mathcal{C}_1$ be the collection of all closed unbounded subsets of $[H_{\theta_1}]^\omega$ consisting of countable elementary submodels M of $(H_{\theta_1}, \in)$ such that $\{T, K\} \subseteq M$. Finally, let $\theta_2 = (2^{2^{\aleph_1}})^+$.

Theorem 66. *Assume* $\mathfrak{mm} > \omega_1$, *and suppose* a_ξ ($\xi \in \Gamma$) *is a given sequence of finite subsets of* T *indexed by some set* $\Gamma \subseteq \omega_1$ *such that* $a_\xi \subseteq T_\xi$ *for all* $\xi \in \Gamma$. *Then there is* $C \in \mathcal{C}_1$ *such that for every* $M \in C$, *if* $\xi = M \cap \omega_1 \in \Gamma$ *then there is* $D \in M \cap \mathcal{C}_0$ *such that every element of* $D \cap M$ *rejects* a_ξ *or no element of* $D \cap M$ *rejects* a_ξ.

Proof. Define $F : [H_{\theta_2}]^\omega \to [H_{\theta_1}]^\omega$ such that for M a countable elementary submodel of H_{θ_2} with $\delta = M \cap \omega_1$ in Γ then $F(M)$ is the collection of all countable elementary submodels N of H_{θ_1} which reject a_δ provided that this set is M-stationary; otherwise, $F(M)$ is the collection of all countable elementary submodels N of H_{θ_1} for which we can find $D \in N \cap \mathcal{C}_0$ such that no $X \in C \cap N$ rejects a_δ. Clearly F is an open and stationary set mapping, so applying the Open Stationary Set-Mapping Reflection (see Chapter 17), we get a continuous $\in$-chain $\langle N_\xi : \xi < \omega_1 \rangle$ of countable elementary submodels N_ξ of $(H_{\theta_2}, \in)$ with the property that if $N_\xi = \delta_\xi$ belongs to Γ then either a every term in a tail of the sequence $\langle N_\eta : \eta < \xi \rangle$ rejects a_{δ_ξ}, or else, for every member N_η in a tail of the sequence $\langle N_\eta : \eta < \xi \rangle$ there $D \in \mathcal{C}_0 \cap N_\eta$ with no member of $D \cap N_\eta$ rejecting a_{δ_ξ}.

The conclusion of the lemma would follow if we show that for every countable elementary submodel M of $(H_{\theta_2}, \in)$ containing all the object accumulated so far and having the property that $\delta = M \cap \omega_1$ belongs to Γ there is $D \in M \cap \mathcal{C}_0$ such that every element of $D \cap M$ rejects a_δ or no element of $D \cap M$ rejects a_δ. We may assume that the second alternative fails, i.e., that for every $D \in \mathcal{C}_0 \cap M$ there is $N \in D \cap M$ rejecting a_δ. It follows that for this M, the set $F(M)$ is the collection of all countable elementary submodels N of H_{θ_1} which reject a_δ and this means that each model in a tail $(N_\xi : \nu_0 < \xi < \delta)$ of the sequence $(N_\xi : \xi < \delta)$ rejects a_δ. Let D_0 be the collection of all countable elementary submodels N of $(H_{\theta_0}, \in)$ such that ν_0 and $\langle N_\xi \cap H_{\theta_0} : \xi < \omega_1 \rangle$ belong to N. Clearly, D_0 is a closed and unbounded subset of $[H_{\theta_0}]^\omega$. We show that every $N \in D_0 \cap M$ rejects a_δ. So, consider an $N \in D_0 \cap M$ and let $\nu = N \cap \omega_1$. Then $\nu \in (\nu_0, \delta)$ and so N_ν rejects a_δ. Since $N_\nu \subseteq N$, we conclude that N also rejects a_δ. This completed the proof. $\qquad \square$

9.4   Coloring axiom for coherent trees

In this section, we prove the following result.

Theorem 67. *Assuming* $\mathfrak{mm} > \omega_1$, *for every coherent tree* T *and every subset* K *of* T *there is an uncountable subset* X *of* T *such that either* $\wedge(X) \subseteq K$ *or* $\wedge(X) \cap K = \emptyset$. $\qquad \square$

Proof. It suffices to concentrate only on our fixed spacial binary coherent tree T. We shall assume that there is no uncountable $X \subseteq T$ such that

$\wedge(X) \subseteq K^c$ and we shall show how to force something similar to the hypothesis of Theorem 65 so that its proof would give us uncountable $X \subseteq T$ such that $\wedge(X) \subseteq K$. There is a natural partial ordering $\mathcal{P}_K$ for achieving this, the collection of all pairs $p = (F_p, \mathcal{N}_p)$ such that:

(1) F_p is a finite subset of T,

(2) $\mathcal{N}_p$ is a finite $\in$-chain of countable elementary submodels of $(H_{\theta_2}, \in)$ containing all the relevant objects,

(3) for every $N \in \mathcal{N}_p$ there is $D \in \mathcal{C}_0 \cap N$ such that F_p is not rejected by any member of $D \cap N$.

We consider $\mathcal{P}_K$ as a poset ordered by coordinatewise inclusions. We shall prove that $\mathcal{P}_K$ is a proper partial order. This will finish the proof, since we are assuming that no countable elementary submodel N of H_{θ_0} rejects a singleton $\{t\}$, any condition of the form $(\{t\}, \{M\})$ with $t \in T_{M \cap \omega_1}$ forces that the union of the first coordinates of the generic filter is uncountable.

Suppose that for some large enough regular cardinal $\theta > \theta_2$ there is some countable elementary submodel M of $(H_\theta, \in)$ containing all relevant objects and there is some $p \in \mathcal{P}_K \cap M$ which cannot be extended to a M-generic condition of $\mathcal{P}_K$. In particular, the condition

$$q = (F_p, \mathcal{N}_p \cup \{M \cap H_{\theta_2}\})$$

is not M-generic. So there is a dense-open subset $\mathcal{D}$ of $\mathcal{P}_K$ such that $\mathcal{D} \in M$ and there is extension r of q such that no $\bar{r} \in \mathcal{D} \cap M$ is compatible with r. Extending r, we may assume that it belongs to $\mathcal{D}$, so our assumption that the standard Fubini-type argument (see, e.g., Chapter 8 above) that tries to build a copy $\bar{r}$ of r that belongs to $\mathcal{D} \in M$ and that is compatible with r must fail. This results into two stationary sets $\mathcal{S} \subseteq [H_{\theta_1}]^\omega$ and $\mathcal{T} \subseteq [H_{\theta_2}]^\omega$ and two mappings $N \mapsto a_N$ and $N \mapsto b_N$ with domains $\mathcal{S}$ and $\mathcal{T}$, respectively, and ranges included in $[T]^{<\omega}$ such that:

(4) for every $N \in \mathcal{S}$ there is a countable elementary submodel M of $(H_{\theta_2}, \in)$ such that $N = M \cap H_{\theta_1}$,

(5) for every $N \in \mathcal{S}$ there is $D \in \mathcal{C}_0 \cap N$ such that no model from $D \cap N$ rejects the finite set a_N,

(6) for every $N \in \mathcal{T}$ there is $D \in \mathcal{C}_0 \cap N$ such that no model from $D \cap N$ rejects the finite set b_N,

(7) for every $N_1 \in \mathcal{T}$ and every $N_0 \in \mathcal{S} \cap N_1$ there is no $D \in \mathcal{C}_0 \cap N_0$ with property that no member of $D \cap N_0$ rejects $a_{N_0} \cup b_{N_1}$.

We may assume that $\mathcal{S} \in N$ for all $N \in \mathcal{T}$. Furthermore, we may assume that for every $N \in \mathcal{T}$, the set b_N has some fixed size n and that $b_N \subseteq T_{M \cap \omega_1}$.

Moreover, applying the pressing down lemma and shrinking $\mathcal{T}$, we may assume that for some fixed $D_0 \in \mathcal{C}_0$ and some fixed countable ordinal ν_0 and an some fixed n-element subset of T_{ν_0}, we have the following conditions on every $N \in \mathcal{T}$,

(8) $D_0 \in N$ and no member of $D_0 \cap N$ rejects b_N,

(9) $s \restriction [\nu_0, N \cap \omega_1) = t \restriction [\nu_0, N \cap \omega_1)$ for all $s, t \in b_N$,

(10) $b_N \restriction \nu_0 = F_0$.

Let U be the collection of all n-element subsets a of T such that for some ordinal $\xi > \nu_0$, the set

$$\mathcal{T}[a] = \{N \in \mathcal{T} : b_N \restriction \xi = a\}$$

is stationary. Note that the natural coordinatewise ordering makes U a tree which is also coherent being isomorphic to a subtree of $\{t \in T : t \restriction \nu_0 = F_0(0)\}$. We shall now see that there is a natural ccc poset which forces an uncountable antichain A of U with the property that for all $a \neq b$ in A there exist $i < n$ such that $a(i) \wedge b(i) \notin K$. Since no member of D_0 rejects any member of A this would contradict Theorem 64, and so the proof of Theorem 67 would be finished.

Let $\mathcal{Q}$ be the collection of all finite antichains q of U that meet a given level of U in at most one point such that

$$\{a(i) \wedge b(i) : i < n\} \not\subseteq K \text{ for all } a \neq b \text{ in } q. \tag{9.4}$$

We order $\mathcal{Q}$ by inclusion and we check that it satisfies the countable chain condition. So, let A be a given uncountable subset of $\mathcal{Q}$. By a Δ-system lemma, we may assume that A forms a Δ-system. Since the root can't contribute to incompatibility between two conditions from A, we may actually assume that it is empty and that in fact A is an uncountable family of pairwise disjoint antichains a of U satisfying the condition (9.4) and all having some fixed size m. We may view an element a of A as a sequence $a(i)$ $(i < m)$ which enumerates a according to the height in U. Since each $a(i)$ is in turn an n-element subset of some level of T, we can order it lexicographically as $a(i)(j) = a(i, j)$ $(j < n)$.

Choose $N \in \mathcal{S}$ such that ν_0 and A are elements of N and let M be a countable elementary submodel of $(H_{\theta_2}, \in)$ such that $N = M \cap H_{\theta_1}$. Let $\delta = N \cap \omega_1$ and fix an element b of A all of whose members are of height $> \delta$. Fix also $C_0 \in \mathcal{C}_0$ such that no member of $C_0 \cap N$ rejects a_N. For each $i < m$, we choose $M_i \in \mathcal{T}$ such that $N \in M_i$ and $b(i)$ is a restriction of b_{M_i}. Since every finite subset of T_δ is equal to the δth member of a ω_1-sequence belonging to M, applying Theorem 66 to the union of a_N and each of the restrictions of b_{M_i} $(i < m)$ to the δth level of T, and then intersecting finitely many closed and unbounded subsets of $[H_{\theta_0}]^\omega$, and then relying on the condition (7), we can find $N_0 \in C_0 \cap N$ such that

$$\bigcap_{i<m} K^c(a_N \cup b(i)) \in \mathcal{U}_{N_0}(T). \tag{9.5}$$

Let $\delta_0 = N_0 \cap \omega_1$. Pick $\nu_1 \in (\nu_0, \delta_0)$ such that elements of $\cup b$ agree on the interval $[\nu_1, \delta_0)$, and moreover, two different nodes of $\cup b$ have different restrictions on δ then they also have different restrictions on ν_1. Using the elementarity of N_0, we can choose a sequence $\langle b_\xi : \xi < \omega_1 \rangle \in N_0$ of elements of A such that $b_{\delta_0} \upharpoonright \delta_0 = b \upharpoonright \delta_0$ and such that for all $\xi < \omega_1$, the nodes of $\cup b_\xi$ are of heights $\geq \xi$, they all agree on the interval $[\nu_1, \xi)$, and if two of nodes of $\cup b_\xi$ have different restrictions on ξ then they also have different restrictions on ν_1.

Using coherence of T, we find uncountable subset Γ of ω_1 in N_0 such that the node $t = b(0,0) \upharpoonright \delta_0$ is in the downwards closure of the set $X = \{b_\xi(0,0) : \xi \in \Gamma\}$, and such that

$$\Delta(t, X) \subseteq \bigcap_{i<m} K^c(a_N \cup b_i). \tag{9.6}$$

Since N_0 belongs to $C_0 \cap N$, it does not reject a_N, and so in particular, $\Delta(t, X)$ is not a subset of $K^c(a_N)$. So there must be $\gamma \in \Gamma \cap N_0$ such that

$$a_N(k) \upharpoonright \Delta(t, b_\gamma(0,0)) \in K \text{ for all } k < |a_N|. \tag{9.7}$$

Using this and (9.6), we conclude that for all $i < m$ there is $j < n$ such that,

$$b(i,j) \upharpoonright \Delta(t, b_\gamma(0,0)) = b(i,j) \wedge b_\gamma(i,j) \notin K. \tag{9.8}$$

We shall show that $b \cup b_\gamma$ satisfies the requirement (9.4) and this will finish the proof. Consider $i_0 \neq i_1 < m$. If $b(i_0) \upharpoonright \delta_0 \neq b(i_1) \upharpoonright \delta_0$ then there is $j < n$ such that $b_\gamma(i_0, j) \wedge b_\gamma(i_1, j) \notin K$. Then $\Delta(b(i_1, j), b_\gamma(i_1, j)) \geq \nu_1 > \Delta(b_\gamma(i_0, j), b_\gamma(i_1, j))$, and so we must have that

$$\Delta(b_\gamma(i_0, j), b(i_1, j)) = \Delta(b_\gamma(i_0, j), b_\gamma(i_1, j)), \tag{9.9}$$

and therefore, $b_\gamma(i_0, j) \wedge b(i_1, j) = b_\gamma(i_0, j) \wedge b_\gamma(i_1, j) \notin K$. If, on the other hand, $b(i_0) \upharpoonright \delta_0 = b(i_1) \upharpoonright \delta_0$, then for all $j < m$,

$$\Delta(b_\gamma(i_0, j), b(i_1, j)) = \Delta(b_\gamma(i_0, j), b(i_0, j)) = \Delta(b_\gamma(0,0), b(0,0)). \tag{9.10}$$

By (9.8), there is $j < m$ such that

$$b_\gamma(i_0, j) \wedge b(i_0, j) = b(i_0, j) \upharpoonright \Delta(b_\gamma(0,0), b(0,0)) \notin K. \tag{9.11}$$

Hence, $b_\gamma(i_0, j) \wedge b(i_1, j) \notin K$. This finishes the proof. $\square$

Chapter 10

Applications to the S-space Problem and the von Neumann Problem

10.1 The S-space problem and its relatives

Theorem 68. *If* $\mathfrak{mm} > \omega_1$ *then every hereditarily separable regular space* X *is hereditarily Lindelöf.*

Proof. If not, there is a subspace Y of X and a well-ordering $<_w$ such that $otp(Y, <_w) = \omega_1$ and for every $y \in Y$, $\{x \in Y : x <_w y\}$ is relatively open in Y. By T_3, for each $y \in Y$ choose U_y, V_y, open in Y, such that $y \in U_y \subseteq \overline{U_y} \subseteq V_y$ and $U_y \cap Y \subseteq \{x \in Y : x \leq_w y\}$.

Define a set-mapping $F : Y \to \mathcal{P}(Y)$ as follows:

$$F(y) = U_y \cap Y.$$

Claim. *The set-mapping F is sparse relative to the σ-ideal of countable subsets of Y.*

Proof. Take $Z \subseteq Y$ uncountable. Let $Z_0 \subseteq Z$ be a countable set such that $\overline{Z_0} \supseteq Z$.

Let $\bar{z} \in Y$ be such that $z <_w \bar{z}$ for all $z \in Z_0$. Take a finite set $p \subseteq \{y \in Y : y >_w \bar{z}\}$ is finite. Then $Z_0 \not\subseteq \bigcup_{y \in p} F(y) = \bigcup_{y \in p} \overline{U_y}$. Otherwise, $\overline{Z_0} \subseteq \bigcup_{y \in p} \overline{U_y}$ which is impossible. ☐

Let $Z \subseteq Y$ be an uncountable F-free set. Then for every $y \neq z \in Z$ we have that $z \notin U_y$. Contradiction. ☐

Theorem 69. *If* $\mathfrak{mm} > \omega_1$ *then every compact ccc* T_5 *space is countably tight.*

Proof. Let X be a compact, ccc, T_5 space. By a previous theorem, X is separable. Suppose X is not countably tight and work for a contradiction. Then, X has a free sequence. We can think of it as a sequence of pairs (F_α, G_α), $\alpha < \omega_1$, such that:

(1) F_α is a G_δ-set, regular-closed.

(2) G_α is a F_σ-set, regular-open.

(3) For all $A, B \in [\omega_1]^{<\omega}$, $A < B$,

$$\left(\bigcap_{\alpha \in A} G_\alpha^c\right) \cap \left(\bigcap_{\alpha \in B} F_\alpha\right) \neq \emptyset.$$

Let D be a countable dense set in X. For a fixed $\xi \in \omega_1$,

$$\{G_\alpha^c \cap D : \alpha < \omega_1\} \cup \{F_\alpha \cap D : \alpha < \omega_1\}$$

is a centered family of infinite subsets of D, so by $\mathfrak{m} > \omega_1$ there is an infinite $c_\xi \subseteq D$ such that:

(4) $c_\xi \subseteq^* F_\beta$ for all $\beta \geq \xi$.

(5) $c_\xi \perp G_\beta$, for all $\beta \leq \xi$.

Recursively on β, choose $a_\beta \subseteq D$ such that

(6) $c_\xi \subseteq^* a_\beta$ for all $\xi \leq \beta$.

(7) $a_\beta \subseteq^* F_\gamma$, for all $\beta \leq \gamma$.

(8) $a_\beta \setminus a_\alpha \subseteq^* G_\alpha$, for all $\beta \leq \alpha$.

Note: To achieve this use $\mathfrak{mm} > \omega_1$, i.e., define a forcing notion which accomplish this. Also there must be a previous lemma implying theses (6), (7) and (8).

We intend to apply SMP. Let

$$\{(\alpha, \beta, \gamma) : \alpha < \beta < \gamma < \omega_1\}$$

and define $F :\to \mathcal{P}(S)$ such that

$$F(\alpha, \beta, \gamma) = \{(\alpha', \beta', \gamma') : \gamma' < \alpha \ \& \ (a_\gamma \setminus a_\beta) \cap (a_{\beta'} \setminus a_{\alpha'}) = \emptyset\}.$$

Notice that $F(\alpha, \beta, \gamma)$ is countable.

Claim 69.1. *F is a sparse set-mapping.*

Proof. Let $\mathcal{I}$ be the σ-ideal of subsets X of S such that $otp(X, <_{lex}) < \omega_1^3 (= otp(S))$. Let $X \in \mathcal{I}^+$. We need to find a countable $E \subseteq X$ and $H \in \mathcal{I}$ such that $E \not\subseteq \bigcup\{F(\alpha, \beta, \gamma) : (\alpha, \beta, \gamma) \in p\}$, for every $p \in [S \setminus H]^\omega$. Shrinking X we may assume that there is an everywhere countably branching tree $T \subseteq [\omega_1]^{\leq 3}$ such that $T^{max} = T \cap [\omega_1]^3 = X$.

Find a countable $E \subseteq X$ such that $\{(a_{\alpha,\beta}, a_\gamma) : (\alpha, \beta, \gamma) \in E\}$ is dense in $\{(a_{\alpha,\beta}, a_\gamma) : (\alpha, \beta, \gamma) \in X\}$ relative to the topology induced from $\mathcal{P}(\mathbb{N})^3$.

Let $\delta = sup\{\gamma : (\alpha, \beta, \gamma) \in E\} < \omega_1$, $H = \{(\alpha, \beta, \gamma) \in S : \alpha \leq \delta\}$.

Notice that $otp(H) < \delta \times \omega_1 \times \omega_1 < \omega_1^3$.

Choose $p \in [S \setminus H]^{<\omega}$. Find $(\alpha', \beta', \gamma') \in T^{max} = X$ such that $\alpha' < \delta < \gamma < \beta' < \gamma'$ for all $(\alpha, \beta, \gamma) \in p$.

Clearly, $(a_\gamma \setminus a_\beta) \cap (a_{\beta'} \setminus a_{\alpha'}) \neq \emptyset$ for all $(\alpha, \beta, \gamma) \in p$.

Let $U = \{(a_\xi, a_\eta, a_\zeta) \in X : \forall(\alpha, \beta, \gamma) \in p \ (a_\eta \setminus a_\xi) \cap (a_\gamma \setminus a_\beta) \neq \emptyset\}$.

By closure of E there is $(\bar{\alpha}, \bar{\beta}, \bar{\gamma}) \in E$ such that $(a_{\bar{\alpha}}, a_{\bar{\beta}}, a_{\bar{\gamma}}) \in U$. This means $(\bar{\alpha}, \bar{\beta}, \bar{\gamma}) \in E \setminus \bigcup\{F(\alpha, \beta, \gamma) : (\alpha, \beta, \gamma) \in p\}$. $\square$

By the dense-set version of SMP, there is an uncountable F-free set Y such that $Y \not\subseteq \{(\alpha, \beta, \gamma) \in S : \alpha \leq \delta\}$, for all $\delta < \omega_1$.

So we may assume that

$$Y = \{(\alpha_\xi, \beta_\xi, \gamma_\xi) : \xi < \omega_1 \text{ such that } \gamma_\xi < \alpha_\eta \text{ for all } \xi < \eta\}$$

and moreover, there is a closed set $\Gamma \subseteq \omega_1$ such that $(\forall \xi)[\alpha_\xi, \gamma_\xi] \cap \Gamma = \emptyset$.

Define

$$A = \bigcup_{\xi < \omega_1} cl(a_{\beta_\xi} \setminus a_{\alpha_\xi}) \setminus (a_{\beta_\xi} \setminus a_{\alpha_\xi}),$$

$$B = \bigcup_{\xi < \omega_1} cl(a_{\gamma_\xi} \setminus a_{\beta_\xi}) \setminus (a_{\gamma_\xi} \setminus a_{\beta_\xi}).$$

Notice that $\bar{A} \cap B = A \cap \bar{B} = \emptyset$. Then the following claim will give the contradiction we are working for:

Claim 69.2. *There are no disjoint open U and V such that $U \supseteq A$ and $V \supseteq B$.*

Proof. Suppose such U and V exist. Then for all $\xi < \omega_1$

$$a_{\beta_\xi} \setminus a_{\alpha_\xi} \subseteq^* U$$

and

$$a_{\gamma_\xi} \setminus a_{\beta_\xi} \subseteq^* V.$$

Find an uncountable $\Delta \subseteq \omega_1$ and an integer n, and take $x, y \subseteq \{0, \ldots, n-1\} = n$ such that

(9) $(a_{\beta_\xi} \setminus a_{\alpha_\xi}) \setminus n \subseteq U$.

(10) $(a_{\gamma_\xi} \setminus a_{\beta_\xi}) \setminus n \subseteq V$.

(11) $(a_{\beta_\xi} \setminus a_{\alpha_\xi}) \cap n = x$.

(12) $(a_{\gamma_\xi} \setminus a_{\beta_\xi}) \cap n = y$.

Pick $\xi < \eta$ in Δ. Then $(\alpha_\xi, \beta_\xi, \gamma_\xi) \in F(\alpha_\eta, \beta_\eta, \gamma_\eta)$. A contradiction.

$\square$

This completes the proof of Theorem 69. $\square$

10.2 The P-ideal dichotomy and a problem of von Neumann

Theorem 70. *If* $\mathfrak{mm} > \omega_1$ *then every complete weakly distributive ccc Boolean algebra supports a positive continuous submeasure.*

Before proving this theorem, let us introduce the following:

Definition. $\mu : \mathcal{B} \to [0, \infty)$ is a *strictly positive submeasure* if:

(a) $a \leq b \to \mu(a) \leq \mu(b)$.

(b) $\mu(a \cup b) \leq \mu(a) + \mu(b)$.

(c) $\mu(a) = 0 \to a = 0$.

μ is *continuous* if it moreover satisfies,

(d) $a_n \downarrow 0 \to \mu(a_n) \downarrow 0$.

Remark. It is easy to see that σ-additive measures are continuous.

Definition. $\mathcal{B}$ is *weakly distributive* if for every matrix $(a_{mn})_{m,n \in \mathbb{N}}$ of elements of $\mathcal{B}$ such that $a_{mn} \downarrow_n 0$ there is a diagonal sequence $(a_{mn(m)})_{m \geq 0}$ such that

$$\bigwedge_k \bigvee_{m \geq k} a_{mn(m)} = 0.$$

Notation. let $\mathcal{B}^+ = \mathcal{B} \setminus \{0\}$ and let $MAX(\mathcal{B})$ denote the collection of all the maximal families of pairwise disjoint elements of $\mathcal{B}^+$ (called "partitions of unity"). For $\mathcal{A}, \mathcal{A}' \in MAX(\mathcal{B})$, write $\mathcal{A} \prec \mathcal{A}'$ if $\mathcal{A}$ refines $\mathcal{A}'$ and $\mathcal{A} \prec_{fin} \mathcal{A}'$ if for every $a \in \mathcal{A}$, $\{b \in \mathcal{A}' : a \wedge b > 0\}$ is finite.

Remark. A ccc complete Boolean algebra $\mathcal{B}$ is weakly distributive iff for every sequence $\{\mathcal{A}_m\}_{m \geq 0} \subseteq MAX(\mathcal{B})$ there exists $\mathcal{A} \in MAX(\mathcal{B})$ such that $\mathcal{A} \prec_{fin} \mathcal{A}_m$ for every $m \in \mathbb{N}$.

Definition. Let $\mathcal{I}_\mathcal{B}$ be the collection of all countable subsets X of $\mathcal{B}^+$ for which we can find a maximal family $\mathcal{A}$ of pairwise disjoint elements of $\mathcal{B}^+$ such that for every $a \in \mathcal{A}$, $\{x \in X : a \wedge x > 0\}$ is finite.

Lemma. *If $\mathcal{B}$ is a ccc and weakly distributive complete Boolean algebra then $\mathcal{I}_\mathcal{B}$ is a P-ideal.*

Proof. Given $\{X_m\}_{m \geq 0} \subseteq \mathcal{I}_\mathcal{B}$, take $\mathcal{A}_m \in MAX(\mathcal{B})$ witnessing $X_m \in \mathcal{I}_\mathcal{B}$, for each m.

Since $\mathcal{B}$ is weakly distributive, find $\mathcal{A} \in MAX(\mathcal{B})$ such that $\mathcal{A} \prec_{fin} \mathcal{A}_m$, for all m. Let $\mathcal{A} = \{a_i : i \in \omega\}$. Then $\mathcal{A}$ witness $X \in \mathcal{I}_\mathcal{B}$, where

$$X = \bigcup (X_m \setminus F_m)$$

and

$$F_m = \{x \in X_m : \exists i \leq m(x \wedge a_i > 0)\}.$$

$\square$

Proof of Theorem 70. By PID, we consider the following two alternatives:

(1) There is an uncountable $T \subseteq \mathcal{B}^+$ such that $[T]^\omega \subseteq \mathcal{I}_\mathcal{B}$, or

(2) There is a countable decomposition $\mathcal{B}^+ = \bigcup_{n=0}^\infty S_n$ such that $S_n \perp \mathcal{I}_\mathcal{B}$, for all n.

We claim that (1) is impossible if $\mathcal{B}$ is complete and ccc. Otherwise, let $T = \{t_\xi : \xi < \omega_1\}$ be as in (1) and define:

$$T_{[\alpha,\beta)} = \bigvee_{\alpha \leq \xi < \beta} t_\xi$$

and

$$T_{[\alpha,\omega_1)} = \bigvee \{T_{[\alpha,\beta)} : \beta < \omega_1\}.$$

By ccc, there is a club $\Gamma \subseteq \omega_1$ such that $\{T_{[\alpha,\beta)} : \alpha, \beta \in \Gamma, \alpha < \beta\}$, is a constant sequence. Let t be the constant value.

Let $\gamma_0 < \gamma_1 < \cdots < \gamma_n < \cdots$ be the first ω elements of Γ, and let

$$X = \{t_\xi : \xi < sup(\gamma_i)_i\} \in [T]^\omega.$$

Claim 70.1. $X \notin \mathcal{I}_\mathcal{B}$.

Proof. Otherwise, we can find a maximal family $\mathcal{A}$ of pairwise disjoint elements of $\mathcal{B}^+$ such that for every $a \in \mathcal{A}$ the set $\{x \in X : a \wedge x > 0\}$ is finite. Consider $a \in \mathcal{A}$ such that $a \wedge t > 0$. For each i there is $\xi_i \in [\gamma_i, \gamma_{i+1})$ such that $t_{\xi_i} = t$.

Then $\{t_{\xi_i} : i \in \mathbb{N}\}$ is a subset of $\{x \in X : a \wedge x > 0\}$, but it is infinite. $\square$

So (2) of the dichotomy holds. Consider now $(\mathcal{B}, \triangle)$. Note that to have a strictly positive continuous submeasure it is sufficient to have a $\triangle$-invariant metric d on $(\mathcal{B}, \triangle)$: for such d,

$$\mu_d(a) = sup\{d(0,b) : b \leq a\}$$

is a strictly positive continuous submeasure.

Conversely, if $\mu : \mathcal{B} \to [0, \infty)$ is a strictly positive continuous submeasure then

$$d_\mu(a,b) = \mu(a \triangle b)$$

is a $\triangle$-invariant metric on $(\mathcal{B}, \triangle)$.

Consider the following 'algebraic' notion of convergence of sequences in $\mathcal{B}$:

$$a_n \to 0 \;\; iff \;\; limsup \; a_n = 0$$

$$a_n \to b \;\; iff \;\; a_n \triangle b \to 0.$$

Let τ_S be the collection of all $U \subseteq \mathcal{B}$ with the property that whenever $a_n \to b$, with $b \in U$, there exists $m \in \mathbb{N}$ such that $a_n \in U$ for all $n \geq m$. Using (2), we show that $(\mathcal{B}, \tau_S)$ is metrizable by a metric which is $\triangle$-invariant and for which $(\mathcal{B}, \triangle, \tau_S)$ is a topological group.

Notation. For $S \subseteq \mathcal{B}$, the closure of S in the topology τ_S is the set

$$\overline{S}^{\tau_S} = \{b : \exists (a_n)_n \subseteq S \;\; (a_n \to b)\}.$$

Claim 70.2. *If statement (2) of the dichotomy holds then there is a sequence $\{U_n\}_n$ of τ_S-open neighborhoods of 0 such that $\bigcap_{n=0}^{\infty} U_n = \{0\}$.*

Proof. Assume that the members of the decomposition $\mathcal{B}^+ = \bigcup_n S_n$ are upwards closed. Let $U_n = \mathcal{B}^+ \setminus \overline{S_n}^{\tau_S}$. Pick a sequence $(a_k)_{k \in \mathbb{N}} \subseteq S_n$ such that $a_k \to 0$. Then $\{a_k : k \in \mathbb{N}\} \perp \mathcal{I}_\mathcal{B}$.

Note that $a_k \to 0$ means $\{a_k : k \in \mathbb{N}\} \in \mathcal{I}_\mathcal{B}$:

$$\bigwedge_l \bigvee_{k \geq l} a_l = 0.$$

For every l, let $c_l = \bigvee_{k \geq l} a_l$. Then

$$\mathcal{A} = \{c_l \setminus c_{l+1} :\in \mathbb{N}\}$$

witness that $\{a_k : k \in \mathbb{N}\}$ belongs to $\mathcal{I}_\mathcal{B}$.

□

Claim 70.3. *For $\{U_n\}_n$ as in Claim 70.2, we actually have that*

$$\bigcap_{n=0}^{\infty} \overline{U_n}^{\tau_S} = \{0\}.$$

Proof. Suppose not and pick $c \neq 0$ with $c \in \bigcap_{n=0}^{\infty} \overline{U_n}^{\tau_S}$. Choose $(a_{nm})_{n,m=0}^{\infty} \subseteq U_n$ such that $a_{nm} \to_n c$. By weakly distributivity, pick a diagonal sequence $(a_{nm(n)})_{n=0}^{\infty}$ such that $a_{nm(n)} \to_n c$, i.e.,

$$limsup \ a_{nm(n)} = c = liminf \ a_{nm(n)}.$$

This means,

$$\bigvee_{k} \bigwedge_{n \geq k} a_{nm(n)}.$$

For every k, let $b_k = \bigwedge_{n \geq k} a_{nm(n)}$. Then $b_k \uparrow c$. Since $c > 0$, pick k such that $b_k > 0$, and $\forall n \geq \overline{k} \ b_k \leq a_{nm(n)} \in U_n$. Then $\forall n \geq k \ b_k \in U_n$, since U_n is closed downwards for each n. Hence $b_k \in \bigcap_{n \geq k} U_n$. This is a contradiction since $b_k > 0$.

□

Claim 70.4. $\vee$ *is continuous at 0.*

Proof. Otherwise we can find a downwards closed neighborhood V of 0 such that for every neighborhood U of 0 there exist $x, y \in U$ such that $x \vee y \notin V$.

Consider $\{U_n\}_n$ as in the previous lemma. Let $V_0 = U_0 \cap V$ and pick $x_0, y_0 \in V_0$ with $x_0 \vee y_0 \notin V$. Given V_n and $x_n, y_n \in V_n$, find $V_{n+1} \subseteq U_{n+1}$ such that $x_n \vee V_{n+1} \subseteq V_n$ and $y_n \vee V_{n+1} \subseteq V_n$. (Notice that the function $x \to x \vee a$ is continuous for every $a \in \mathcal{B}$).

Find $x_{n+1}, y_{n+1} \in V_{n+1}$ such that $x_{n+1}, y_{n+1} \notin V$, and let

$$x = limsup \ x_n$$

and

$$y = limsup \ y_n.$$

We finish the proof of this claim by showing

$$x = y = 0.$$

For every k, let $c_k = \bigvee_{n \geq k} x_n$. It is sufficient to show that $c_k \in \overline{U_n}^{\tau_S}$ for all k. Then $x = \bigwedge_k c_k \in \overline{U_n}^{\tau_S}$ and therefore $x = 0$ because $\bigcap_{n=0}^{\infty} \overline{U_n}^{\tau_S} = \{0\}$.

So let
$$c_l = sup\{x_l \vee x_{l+1} \vee \ldots x_k \vee x_{k+1} : k \in \mathbb{N}\}.$$

But $x_l \vee x_{l+1} \vee \ldots x_k \vee x_{k+1} = x_l \vee (x_{l+1} \vee \ldots x_k \vee x_{k+1}) \in V_l \subseteq U_l$, by induction, therefore c_l being the supremum is in $\overline{U_n}^{\tau_S}$. $\qquad \square$

Claim 70.5. $(\mathcal{B}, \tau_S)$ *is* T_3.

Proof. Pick an open V neighborhood of 0. By continuity of $\vee$, there is a downwards closed neighborhood of 0, U, such that $U \vee U = U \triangle U \subseteq V$. Since $\overline{U_n}^{\tau_S} \subseteq U \triangle U$ (this a property of the group), we get $\overline{U_n}^{\tau_S} \subseteq V$. $\qquad \square$

Claim 70.6. $\{U_n\}_n$ *as above is a countable neighborhood base of 0 in* $(\mathcal{B}, \tau_S)$.

Proof. Otherwise, there is a closed neighborhood V of 0 such that $U_n \not\subseteq V$, foe all n. We assume that V is downwards closed. Construct sequences $(x_n)_n \subseteq \mathcal{B}^+$ and as follows:

Find $x_0 \in U_0 \setminus V$ such that there is an open neighborhood V_0 of 0 with $(x_0 \vee V_0) \cap V = \emptyset$ (by the upwards closure of V^c). In the same way, find $x_n \in U_n \setminus V$ and $V_n \subseteq U_n$ with $(x_n \vee V_n) \cap V = \emptyset$.

Then by induction on the difference we show that $x_n \vee x_{n+1} \vee \cdots \vee x_m \in V_n \setminus V$, for every $n < m$ (as in the previous lemma). It follows that $\bigvee_{k \geq n} x_k \notin int(V)$, and therefore $x = \bigwedge_n \bigvee_{k \geq n} x_k \notin int(V)$. But $x = 0$. $\qquad \square$

This completes the proof of Theorem 70. $\qquad \square$

Corollary. *The P-ideal Dichothomy implies that Souslin Hypothesis holds.*

Proof. If SH fails then there is a tree T of height ω_1 with no uncountable chain nor antichain. By removing some points of T we may assume that for every $t \in T$ the set $T^t = \{s \in T : t \leq_T s\}$ of successors of t in T is uncountable. Let τ be the topology on T generated by the family T^t $(t \in T)$ as a basis. Let $\mathcal{B} = ro(T)$ be the algebra of regular-open sets of (T, τ).

Claim. $\mathcal{B}$ *is a ccc weakly distributive algebra.*

Proof. Let $\mathcal{A}_n$ $(n \in \omega)$ be a sequence of maximal antichains of $\mathcal{B}$. For each n, let
$$\mathcal{D}_n = \{t \in T : (\exists a \in \mathcal{A}_n) \ T^t \subseteq a\}.$$

Then each $\mathcal{D}_n$ is dense-open in T and therefore the set A_n of minimal elements of $\mathcal{D}_n$ ia a maximal antichain of T. Let $\alpha < \omega_1$ be such that
$$(\forall \in \bigcup_{n=0}^{\infty} A_n) \ \alpha > ht_T(t).$$

Let $T_\alpha = \{t \in T : ht_T(t) = \alpha\}$. Then T_α is the αth level of T and so is countable. Let

$$B = \{T^t : t \in T_\alpha\}.$$

Then B is a maximal antichain of B which refines all $\mathcal{A}_n$ $(n \in \omega)$. □

So by the proof of Theorem 70, there is a strictly positive continuous submeasure

$$\mu : \mathcal{B} \to [0, 1].$$

Then for some $n \in \omega$, the set

$$X_n = \{t \in T : \mu(T^t) > 2^{-n}\}$$

is uncountable. Let

$$\mathcal{D} = \{s \in T : (\exists t \in X_n) \ s \geq t \ \text{ or } \ (\forall t \in X_n) \ s \perp t\}.$$

Then $\mathcal{D}$ is a dense open subset of T. Let A be the set of all minimal elements of $\mathcal{D}$. Than A is countable, being an antichain of T. Pick $\alpha \in \omega_1$ such that $\alpha \geq ht_T(s)$ for all $s \in A$. Since X_n is uncountable there is $t \in X_n$ of height α. And since A is a maximal antichain, there is $s \in A$ suxh that $s \leq_T t$. It follows that X_n is dense in T^s. So in particular X_n contains an antichain B of size 2^n. It follows that

$$1 \geq \mu(\bigcup_{t \in B} T^t) = \sum_{t \in B} \mu(T^t) > |B| \cdot 2^n > 1,$$

a contradiction.

□

Chapter 11

Biorthogonal Systems

11.1 The quotient problem

Suppose that X is a normed space and let X^* denotes its dual space, the collection of all continuous linear functionals on X with the norm $\|f\|^* = \sup_{x \in X} |f(x)|$. Recall also that a complete normed space is called a Banach space.

Recall that, for a given ordinal γ, a sequence $x_\alpha (\alpha < \gamma)$ of elements of some Banach space X is a transfinite Schauder basis of X if every element x of X has a unique representation as a sum $x = \sum_{\alpha < \gamma} a_\alpha x_\alpha$ (see [24; p. 581]). This is equivalent to the condition that $\overline{\text{span}}\{x_\alpha : \alpha < \delta\} = X$ and that there is of an absolute constant C such that for all $\beta < \gamma$ the projection map $P_\beta : X \longrightarrow \overline{\text{span}}\{x_\alpha : \alpha < \gamma\}$ has norm $\leq C$. Note that the corresponding sequence $x_\alpha^*(\alpha < \gamma)$ of bi-orthogonal functionals is also a transfinite basic sequence in X^* with the same basis constant C though its norm-closed linear span may not be equal to the whole of X^*. What is true however is that every element x^* of X^* has a unique representation as a w^*-convergent series $\sum_{\alpha < \gamma} x^*(x_\alpha) x_\alpha^*$. (see [24; p. 582]). Thus, a convenient way of representing a given Banach space X is via a Schauder basis, short or long. Unfortunately this is not always possible and to get a Schauder basis one is forced to go either to a closed subspace of X or to a quotient space of X. This has been realized implicitly already in Banach's original monograph though the formulations of the basis problem for Banach spaces were fully achieved only some decades later.

Problem 1. Does every infinite dimensional Banach space have an infinite dimensional quotient with a Schauder basis?

The separable case of Problem 1 was solved by Johnson and Rosenthal but in general the problem (better known in the literature in its reformulation as the 'separable quotient problem') is still open. Any proper analysis of Problem 1 involves study of the existence of fundamental biorthogonal systems in X. Recall that family $\{(x_i, f_i) : i \in I\} \subseteq X \times X^*$ forms a *biorthogonal system* whenever $\sup_{i \in I} \|x_i\| < \infty$ and $\sup_{i \in I} \|f_i\|^* < \infty$ and $f_i(x_i) = 1$ and $f_i(x_j) = 0$ and $i \neq j$. Recall also that a biorthogonal system $(x_i, x_i^*)_{i \in I}$ in some Banach space X is said to be *fundamental* if $\overline{\text{span}}\,\{x_i : i \in I\} = X$, *total* if $\bigcap_{i \in I} \ker(x_i^*) = \{0\}$. It is well known (see [24; p. 603]) that every Banach space has a bounded total biorthogonal system. It is also well-known (see [16; pp. 42-47]) that every separable Banach space X admits a total and fundamental biorthogonal system bounded by $1 + \epsilon$. Davis and Johnson have shown that if a Banach space X has a weakly compactly generated quotient of the same density, then X has a bounded fundamental biorthogonal system and have asked the following question:

Problem 2.   Does every Banach space have a bounded fundamental biorthogonal system?

According to a result of Plichko, Problem 2 has the following equivalent formulation which should be compared with Problem 1 above.

Problem 2*.   Does every Banach space have a quotient with a Schauder basis of length equal to its density?

Unfortunately, this question in its full generality has a negative answer; the space $l_\infty^c(\Gamma)$ for an index set Γ of cardinality greater than continuum being a counterexample. So one is immediately led to consider either a restriction on the density of a given Banach space X or a restriction on the nature of X. The following appears as good test question towards Problem 2.

Problem 3.   Does every nonseparable Banach space admit an uncountable biorthogonal system?

It is this question that we answer in this Chapter. It is interesting that in order to solve this problem, we will need to go through the analysis of the quotient problem. In particular, we will need the following result.

Theorem 71. *Suppose X is a Banach space of density $< \mathfrak{m}$ admitting a bounded linear operator $H : X \longrightarrow c_0(\omega_1)$ with nonseparable range. Then X admits a quotient with a Schauder basis of length ω_1.*

Proof: Note that the assumption is equivalent to the existence of a normalized sequence f_α $(\alpha < \omega_1)$ of bounded linear functionals on X such that $\langle f_\alpha(x) : \alpha < \omega_1 \rangle \in c_0(\omega_1)$ for all $x \in X$. We first work under the assumption that the linear subspace $Y = \{x \in X : \sum_{\alpha < \omega_1} |f_\alpha(x)| < \infty\}$ is norm-dense in X and we fix for each $x \in Y$ and $\epsilon > 0$ a finite set $\Gamma_\epsilon(x) \subseteq \omega_1$ such that

$$\sum_{\alpha \notin \Gamma_\epsilon(x)} |f_\alpha(x)| < \epsilon.$$

Our intention is to use Baire category method to select an uncountable transfinite Schauder basic subsequence $f_\alpha (\alpha \in \Gamma)$ in such a way that there is a quotient map from X onto the norm-closed linear span of the corresponding sequence $f_\alpha^* (\alpha \in \Gamma)$ of biorthogonal functionals. This use of the Baire category method is done via the following set of $\mathcal{P}$ of 'finite approximations' to such a subsequence $f_\alpha (\alpha \in \Gamma)$. The elements of $\mathcal{P}$ will be triples

$$p = (D_p, \Gamma_p, \epsilon_p),$$

where D_p is a finite subset of Y, Γ_p is a finite subset of ω_1 and ϵ_p is a rational number from the interval $(0,1)$ such that the following conditions are satisfied:

(1) For every $f^* \in (\text{span}\{f_\gamma : \gamma \in \Gamma_p\})^*$, $\|f^*\| = 1$, there is $x \in D_p$, $\|x\| = 1$ such that $|f^*(e) - e(x)| \leq (\epsilon_p/3) \cdot \|e\|$ for all $e \in \text{span }\{f_\gamma : \gamma \in \Gamma_p\}$.

We order $\mathcal{P}$ by letting $p \leq q$ if $D_p \subseteq D_q$, $\Gamma_p \subseteq \Gamma_q$, $\epsilon_p \geq \epsilon_q$, and

(2) $(\forall x \in D_p)\ \Gamma_{\epsilon_p/3}(x) \cap (\Gamma_q \backslash \Gamma_p) = \emptyset$.

Claim 1. $\mathcal{P}$ satisfies the countable chain condition.

Proof: Let $p_\xi (\xi \in \omega_1)$ be a given sequence of elements of $\mathcal{P}$. We need to find $\xi \neq \eta$ such that p_ξ and p_η admit a common extension. Applying the standard Δ-system lemma and going to a subsequence we may assume to have $\epsilon \in \mathbb{Q} \cap (0,1)$ and finite subsets D and Δ of X and ω_1, respectively such that for all $\xi < \eta < \omega_1$:

(4) $\epsilon_{p_\xi} = \epsilon$,
(5) $D_{p_\xi} \cap D_{p_\eta} = D$,
(6) $\Gamma_{p_\xi} \cap \Gamma_{p_\eta} = \Gamma$,
(7) $\Gamma_{p_\xi} \backslash \Delta < \Gamma_{p_\eta} \backslash \Delta$,
(8) $(\forall x \in D_{p_\xi})(\forall \alpha \in \Gamma_{p_\eta} \backslash \Delta)\ f_\alpha(x) = 0$.

Fix $\eta \in \Gamma$ such that $\Gamma \restriction \eta = \{\xi \in \Gamma \colon \xi < \eta\}$ is infinite. Then there must be $\xi \in \Gamma \restriction \eta$ such that

(9) $(\forall x \in D_{p_\eta})\ \Gamma_\epsilon(x) \cap (\Gamma_{p_\xi} \backslash \Delta) = \emptyset$.

Let $r = (D_r, \Gamma_{p_\xi} \cup \Gamma_{p_\eta}, \epsilon)$, where $D_r \supseteq D_{p_\xi} \bigcup D_{p_\eta}$ is any finite subset of Y that ensures that r satisfies (1) and therefore that it belongs to $\mathcal{P}$. The existence of such a subset follows, for example, from the principle of local reflexivity. Then by (7) and (8) we conclude that $p_\xi \leq r$ and $p_\eta \leq r$, as required. $\square$

For $\delta \in \omega_1$ let

$$\mathcal{D}_\gamma^0 = \{p \in \mathcal{P} : (\exists \alpha \in \Gamma_p)\alpha \geq \gamma\}.$$

Claim 2. $(\forall p \in \mathcal{P})(\forall \gamma \in \omega_1)(\exists q \in \mathcal{D}_\gamma^0)\ p \leq q$.

Proof: Given $p \in \mathcal{P}$ and $\delta \in \omega_1$, since $\langle f_\alpha(x) : \alpha < \omega_1\rangle \in C_0(\omega_1)$ for each $x \in D_p$, we can find $\alpha \geq \gamma$ such that $f_\alpha(x) = 0$ for all $x \in D_p$. Using the principle of local reflexivity again, we find a finite subset D_q of Y containing D_p ensuring that

$$q = (D_q, \Gamma_p \cup \{\alpha\}, \epsilon_p),$$

satisfies the conditions (1) and that it therefore belongs to $\mathcal{P}$. It is clear that then actually $q \in \mathcal{D}_\gamma^0$ and $p \leq q$.   □

For $x \in X$, let

$$\mathcal{D}_x^1 = \{p \in \mathcal{P} : x \in D_p\}.$$

Claim 3. $(\forall x \in Y)(\forall p \in \mathcal{P})(\exists q \in \mathcal{D}_x^1) p \leq q.$

Proof: Set $q = (D_p \cup \{x\}, \Gamma_p, \epsilon_p)$.   □

For $\epsilon \in \mathbb{Q} \cap (0, 1)$, let $\mathcal{D}_\epsilon^2 = \{p \in \mathcal{P} : \epsilon_p \leq \epsilon\}.$

Claim 4. $(\forall \epsilon \in \mathbb{Q} \cap (0, 1))(\forall p \in \mathcal{P})(\exists q \in \mathcal{D}_\epsilon^2) \ p \leq q.$

Proof: Given ϵ and p, set $q = (D_q, \Gamma_p, \min\{\epsilon_p, \epsilon\})$, where $D_q \supseteq D_p$ is any finite subset of Y ensuring that q satisfies the condition (1).   □

Let Y_0 be a norm-dense subset of Y of size equal to the density of X. Since by our assumption $\omega_1 \leq$ dens $(X) < \mathfrak{m}$, there is a filter $\mathcal{F} \subseteq \mathcal{P}$ such that:

(10) $(\forall \alpha < \omega_1) \ \mathcal{F} \cap \mathcal{D}_\alpha^0 \neq \emptyset,$

(11) $(\forall x \in \operatorname{span}_\mathbb{Q} Y_0) \ \mathcal{F} \cap \mathcal{D}_x^1 \neq \emptyset,$

(12) $(\forall \epsilon \in \mathbb{Q} \cap (0, 1)) \ \mathcal{F} \cap \mathcal{D}_\epsilon^2 \neq \emptyset.$

Let $\Gamma_0 = \bigcup_{p \in \mathcal{F}} \Gamma_p$. Then Γ_0 is an uncountable subset of ω_1. Let Ω be the collection of all limit ordinals $\delta < \omega_1$ such that:

(13) $(\forall \epsilon \in \mathbb{Q} \cap (0, 1))(\forall \Delta \in [\Gamma]^{<\omega} \upharpoonright \delta)(\exists p \in \mathcal{F})[\epsilon_p \leq \epsilon \ \& \ \Delta \subseteq \Gamma_p \subseteq \delta].$

Then Ω is a closed and unbounded subset at ω_1. Let

$$\Gamma = \{\beta \in \Gamma_0 : (\forall \alpha < \beta)(\alpha \in \Gamma_0 \to \Omega \cap (\alpha, \beta] \neq \emptyset)\}.$$

Thus Γ_0 is a natural uncountable subset of Γ which has the property that between any two consecutive elements $\gamma < \delta$ of Ω there is at most one element of Γ. Going to an uncountable subset of Γ we may assume that $f_\gamma(\gamma \in \Gamma)$ is linearly independent.

For $\delta \in \Omega$ let

$$P_\delta : \operatorname{span}\{f_\alpha : \alpha \in \Gamma\} \to \operatorname{span}\{f_\alpha : \alpha \in \Gamma \upharpoonright \delta\}$$

be the natural projection.

Claim 5. $(\forall \delta \in \Omega)(\forall f \in \operatorname{span}\{f_\alpha : \alpha \in \Gamma\}) \ \|P_\delta(f)\| \leq \|f\|.$

Proof: Otherwise, we can find $f = \sum_{\alpha \in \Delta} a_\alpha f_\alpha$ in $\operatorname{span}\{f_\alpha : \alpha \in \Gamma\}$ such that if $g = P_\delta(f) = \sum_{\alpha \in \Delta \upharpoonright \delta} a_\alpha f_\alpha$ then $\|g\| > (1 + \epsilon)\|f\|$ for some $\epsilon > 0$.

Multiplying f by a positive scalar, we may (and will) assume that $\|g\| = 1$. Let $a = 1 + \max_{\alpha \in \Delta} |a_\alpha|$. Since $\delta \in \Omega$, we can find $p \in \mathcal{F}$ such that:

(14) $\epsilon_p < \epsilon/4 \cdot a$,

(15) $\Delta \upharpoonright \delta \subseteq D_p \subseteq \{\alpha : \alpha < \delta\}$.

Find $q \in \mathcal{F}$ such that $\Delta \subseteq \Gamma_q$ and $r \in \mathcal{F}$ such that $p \leq r$ and $q \leq r$. Since p satisfies (1) we can find $x \in D_p$ such that

(16) $|g(x)| > 1 - \epsilon_p/3$.

From $p \leq r$ and the fact that $\Gamma_r \supseteq \Gamma_q \supseteq \Gamma$, we infer (see (2))

(17) $\sum_{\alpha \in \Delta^\delta} |f_\alpha(x)| < \epsilon_p/3$,

where $\Delta^\delta = \{\alpha \in D : \alpha \geq \delta\}$. It follows that

$$\begin{aligned}
1 = \|g\| &\geq (1 + \epsilon)\|f\| \geq (1 + \epsilon)|f(x)| \\
&\geq (1 + \epsilon)(|g(x)| - |\sum_{\alpha \in \Delta^\delta} a_\alpha f_\alpha(x)|) \\
&\geq (1 + \epsilon)(1 - \epsilon_p/3 - a \cdot \sum_{\alpha \in \Delta^\delta} |f_\alpha(x)|) \\
&\geq (1 + \epsilon)(1 - \epsilon_p/3 - a \cdot \epsilon_p/3) > 1,
\end{aligned}$$

a contradiction. □

It follows that $f_\alpha (\alpha \in \Gamma)$ is a transfinite Schauder basic sequence with basic constant 1. For a subset $\Delta \subseteq \Gamma$ let

$$P_\Delta : \overline{\text{span}}\{f_\alpha : \alpha \in \Gamma\} \to \overline{\text{span}}\{f_\alpha : \alpha \in \Delta\}$$

be the projection.

Claim 6. $(\forall \epsilon \in (0,1))(\forall p \in \mathcal{F})\, (\epsilon_p < \epsilon/2 \to \|P_{\Gamma_p \cap \Gamma}\| \leq 1 + \epsilon)$.

Proof: Otherwise, we can find $f = \sum_{\alpha \in \Delta} a_\alpha f_\alpha$ in $\text{span}\{f_\alpha : \alpha \in \Gamma\}$ of norm 1 such that, if we let $g = P_{\Gamma \cap \Gamma_p}(f) = \sum_{\alpha \in \Delta \cap \Gamma_p} a_\alpha f_\alpha$, then $\|g\| > (1 + \epsilon)\|f\| = 1 + \epsilon$. Let $\lambda = \|g\|$. By (1) we can find $x \in D_p$ such that

(18) $|(1/\lambda)g(x)| > 1 - \epsilon_p/3$.

Then

$$\frac{1}{1+\epsilon} > \|(1/\lambda)\cdot f\| \ge |(1/\lambda)f(x)|$$

$$\ge |(1/\lambda)g(x)| - (\frac{1}{\lambda})|\sum_{\alpha\in\Delta\setminus\Gamma_p} a_\alpha f_\alpha(x)|$$

$$\ge 1 - \epsilon_p/3 - \frac{2}{\lambda}\sum_{\alpha\in\Delta\cap\Gamma_p}|f_\alpha(x)|$$

$$\ge 1 - \epsilon_p/3 - 2\epsilon_p/3 = 1 - \epsilon_p > \frac{1}{1+\epsilon},$$

a contradiction.   □

Let $\{f_\gamma^* : \gamma \in \Gamma\} \subseteq (\overline{\text{span}}\{f_\gamma : \gamma \in \Gamma\})^*$ be the sequence of functionals biothorgonal to $\{f_\gamma : \gamma \in \Gamma\}$. Then $f_\gamma^*(\gamma \in \Gamma)$ is also a transfinite Schauder basic sequence with basic constant 1.

Let

$$T = I^* : X \to (\overline{\text{span}}\{f_\gamma : \gamma \in \Gamma\})^* (= \overline{\text{span}}^*\{f_\alpha^* : \alpha \in \Gamma\})$$

be the natural restriction operator, a bounded linear operator dual to the inclusion operator

$$I : \overline{\text{span}}\{f_\alpha : \alpha \in \Gamma\} \to X^*.$$

Note also that for every $x \in \text{span}_\mathbb{Q} X_0$,

$$T(x)(f) = f(x) = \sum_{\alpha\in\Gamma} f_\alpha(x)\cdot f_\alpha^*$$

is an absolutely converging series in $(\overline{\text{span}}\{f_\gamma : \gamma \in \Gamma\})^*$.

Claim 7. $T[X] = \overline{\text{span}}\{f_\delta^* : \delta \in \Gamma\}$.

Proof: Consider an $x \in \text{span}_\mathbb{Q} Y_0$. Then there is $p \in \mathcal{F}$ such that $x \in D_p$ and therefore $\sum_{\alpha\in\Gamma} |f_\alpha(x)| < \infty$. It follows that $T(x) = \sum_{\alpha\in\Gamma} f_\alpha(x).f_\alpha^* \in \overline{\text{span}}\{f_\gamma : \gamma \in \Gamma\}$. This shows that $T[\text{span}_\mathbb{Q} X_0] \subseteq \overline{\text{span}}\{f_\gamma : \gamma \in \Gamma\}$. Since T is a bounded operator, this gives us the first inclusion $T[X] \subseteq \overline{\text{span}}\{f_\delta : \delta \in \Gamma\}$. The reverse inclusion will follow from the following fact via a standard argument of successive approximation.

Subclaim 8. $(\forall\epsilon \in (0,1))(\forall g^* \in \text{span}\{f_\alpha^* : \alpha \in \Gamma\}), \|g^*\| = 1, (\exists x \in X), \|x\| = 1, \|T(x) - g^*\| \le 4\epsilon.$

Proof: Let $\epsilon \in (0,1)$ and $g^* = \sum_{\alpha\in\Delta} a_\alpha f_\alpha^* \in \text{span}\{f_\gamma^* : \gamma \in \Gamma\}$ with $\|g^*\| = 1$ be given. Find $p \in \mathcal{F}$ such that $\epsilon_p < \epsilon/2$ and $\Gamma_p \supseteq \Delta$. For $f^* \in \text{span}\{f_\gamma^* : \gamma \in \Gamma_p \cap \Gamma\}$, set

$$\|f^*\| = \sup\{|f^*(e)| : e \in \text{span}\{f_\gamma : \gamma \in \Gamma_p\cap\Gamma\}, \|e\| \le 1\}.$$

Then for every $f^* \in \text{span}\{f_\gamma^* : \gamma \in \Gamma_p \cap \Gamma\}$,

(19) $\||f^*\|| \leq \|f^*\| \leq \|P_{\Gamma \cap \Gamma_p}\| \cdot \||f^*\|| \leq (1+\epsilon)\||f^*\||.$

To see this, note that $f^* = \sum_{\gamma \in \Gamma_p \cap \Gamma} f^*(f_\gamma)f_\gamma^*$, and therefore

$$\|f^*\| = \sup\{| \sum_{\gamma \in \Gamma_p \cap \Gamma} f^*(f_\gamma) \cdot f_\gamma^*(e)| : e \in \overline{\text{span}}\{f_\gamma : \gamma \in \Gamma\}, \|e\| \leq 1\}$$

$$= \sup\{|f^*(\sum_{\gamma \in \Gamma_p \cap \Gamma} f_\gamma^*(e) \cdot f_\gamma)| : e \in \overline{\text{span}}\{f_\gamma : \gamma \in \Gamma\}, \|e\| \leq 1\}$$

$$= \sup\{|f^*(P_{\Gamma_p \cap \Gamma}(e))| : e \in \overline{\text{span}}\{f_\gamma : \gamma \in \Gamma\}, \|e\| \leq 1\}$$

$$\leq \||f^*\|| \cdot \|P_{\Gamma_p \cap \Gamma}\| \leq \||f^*\|| \cdot (1+\epsilon)$$

by Claim 6.

Let $h^* = g^*/\|g^*\|$. Note that by (19), $\||g^* - h^*\|| \leq \epsilon$. Applying (1) for h^* we find $x \in D_p$ such that

(20) $\||h^* - \sum_{\alpha \in \Gamma_p \cap \Gamma} f_\alpha(x).f_\alpha^*\|| < \epsilon_p/3.$ By (19) and (20), we get that

(21) $\|h^* - \sum_{\alpha \in \Gamma_p \cap \Gamma} f_\alpha(x) \cdot f_\alpha^*\| \leq 2 \cdot (\epsilon_p/3).$

It follows that,

$$\|g^* - T(x)\| \leq \|h^* - T(x)\| + \|g^* - h^*\|$$

$$\leq \|h^* - \sum_{\alpha \in \Gamma_p \cap \Gamma} f_\alpha(x) \cdot f_\alpha^*\| + \| \sum_{\alpha \in \Gamma \setminus \Gamma_p} f_\alpha(x) \cdot f_\alpha^*\| + \epsilon$$

$$\leq (2/3)\epsilon_p + 2 \cdot \sum_{\alpha \in \Gamma \setminus \Gamma_p} | f_\alpha(x) | + \epsilon$$

$$\leq (2/3)\epsilon_p + 2 \cdot \epsilon_p + \epsilon \leq 4\epsilon,$$

as required.

The proof of Theorem 71 is finished once we show that under its hypothesis, there is always an uncountable $\Gamma \subseteq \omega_1$ such that

$$\{x \in X : \sum_{\gamma \in \Gamma} |f_\gamma(x)| < \infty\}$$

is norm-dense in X. This will be done again using a natural poset $\mathcal{P}_0$ of finite approximations to such a set Γ. The members of $\mathcal{P}_0$ are pairs $p = (D_p, \Gamma_p)$ where D_p is a finite subset of X, where Γ_p is a finite subset of ω_1, and where, for all $\gamma \in D_p$,

(22) $\sum_{\gamma \in \Gamma_p} |f_\gamma(x)| < 1.$

We order $\mathcal{P}_0$ by letting $p \leq q$ if $D_p \subseteq D_q$ and $\Gamma_p \subseteq \Gamma_q$.

Claim 9. $\mathcal{P}_0$ satisfies the countable chain condition.

Proof: Let $p_\xi(\xi < \omega_1)$ be a given sequence of members of $\mathcal{P}_0$. Going to an uncountable subsequence we may assume that for some $\epsilon > 0$ and all $\xi < \omega_1$.

(23) $(\forall x \in D_{p_\xi}) \sum_{\gamma \in \Gamma_p} |f_\gamma(x)| < 1 - \epsilon.$

Using the Δ-system lemma and thinning the sequence to an uncountable subsequence $p_\xi(\xi \in \Omega)$, we may assume that for some integer k and finite sets $D \subseteq X$ and $D \subseteq \omega_1$, we have the following for all $\xi < \eta$ in Ω:

(24) $D_{p_\xi} \cap D_{p_\eta} = D,$
(25) $\Gamma_{p_\xi} \cap \Gamma_{p_\eta} = \Delta,$
(26) $|\Gamma_{p_\xi} \backslash \Delta| = k,$
(27) $\Gamma_{p_\xi} \backslash \Delta < \Gamma_{p_\eta} \backslash \Delta,$
(28) $(\forall x \in D_{p_\xi})(\forall \gamma \in \Gamma_{p_\eta} \backslash \Delta)\ f_\gamma(x) = 0,$

Pick $\eta \in \Omega$ such that $\Omega \upharpoonright \eta = \{\xi \in \Omega : \xi < \eta\}$ is infinite. By our assumption that $\langle f_\gamma(x) : \gamma < \omega_1 \rangle \in c_0(\omega_1)$ for all $x \in X$, there must be $\xi \in \Omega \upharpoonright \eta$ such that

(29) $(\forall x \in D_{p_\eta})(\forall \gamma \in \Gamma_{p_\xi} \cap \Delta)|f_\gamma(x)| < \epsilon/k.$

Then $q = (D_{p_\xi} \cup D_{p_\eta}, \ \Gamma_{p_\xi} \cup \Gamma_{p_\eta})$ satisfies (22) and so it belongs to $\mathcal{P}_0$. Since clearly $p_\xi \leq q$ and $p_\eta \leq q$ this finishes the proof. $\square$

For $\alpha < \omega_1$, let
$\mathcal{D}_\alpha^0 = \{q \in \mathcal{P}_0 : (\exists \gamma \geq \alpha)\gamma \in \Gamma_q\}.$

Claim 10. $\mathcal{D}_\alpha^0$ is a dense subset of $\mathcal{P}_0$ for all $\alpha < \omega_1$.

Proof: Given $p \in \mathcal{P}_0$, by our assumption that $\langle f_\gamma(x) : \gamma < \omega_1 \rangle \in c_0(\omega_1)$ for all $x \in X$, there is $\beta < \omega_1$ such that $\beta > \alpha$ and

(30) $(\forall x \in D_p)(\forall \gamma \geq \beta)\ f_\gamma(x) = 0.$

Then $q = (D_p, \Gamma_p \cup \{\beta\})$ is a member of $\mathcal{D}_\alpha^0$ extending p. $\square$

Find a norm-dense subset Z of X of size $< \mathfrak{m}$. For $z \in Z$, let

$$\mathcal{D}_z^1 = \{q \in \mathcal{P}_0 : (\exists \lambda > 0)\lambda z \in D_q\}.$$

Claim 11. $\mathcal{D}_z^1$ is a dense subset of $\mathcal{P}_0$ for all $z \in Z$.

Proof: Given $p \in \mathcal{P}_0$ choose an integer k such that

$$k > \sum_{\gamma \in \Gamma_p} |f_\gamma(z)|.$$

Then $q = (D_p \cup \{(1/k)z\}, \Gamma_p)$ is a member of $\mathcal{D}_z^1$ extending p. $\quad\square$

From the definition of $\mathfrak{m}$ there is a filter $\mathcal{F} \subseteq \mathcal{P}_0$ such that:

(31) $(\forall \alpha < \omega_1)$ $\mathcal{F} \cap \mathcal{D}_\alpha^0 \neq \emptyset$,

(32) $(\forall z \in Z)$ $\mathcal{F} \cap \mathcal{D}_\alpha^1 \neq \emptyset$.

Let
$$\Gamma = \bigcup_{p \in \mathcal{F}} \Gamma_p \text{ and } D = \bigcup_{p \in \mathcal{F}} D_p.$$

Then Γ is an uncountable subset of ω_1 and

(33) $(\forall x \in D) \sum_{\gamma \in \Gamma} |f_\gamma(x)| \leq 1$.

Then
$$Y = \{y \in X : \sum_{\gamma \in \Gamma} |f_\gamma(x)| < \infty\}$$

is a linear subspace of X which includes D and therefore Z by (32). It follows that Y is a norm-dense subset of X, which finishes the proof of Theorem 71. $\quad\square$

In the next section we shall show that an arbitrary nonseparable Banach space of density $< \mathfrak{mm}$ satisfies the hypothesis of Theorem 71. For this it will be convenient to split this class of Banach spaces into two subclasses according whether their dual balls equipped with the weak* topology are countably tight or not.

11.2 A topological property of the dual ball

In this section we deal with a class of Banach spaces X such that (B_X^*, w^*) is a countably determined compactum. Recall that, for an index-set Γ, we use $l_\infty^c(\Gamma)$ to denote the closed subspace of $l_\infty(\Gamma)$ consisting of vectors of countable support.

Theorem 72. *Let X be a Banach space of density $< \mathfrak{m}$ whose dual ball equipped with the weak* topology is countably tight. Suppose that there is a sequence $f_\alpha(\alpha < \omega_1)$ of norm-one functionals on X such that for all $x \in X$ the set $\{\alpha < \omega_1 : f_\alpha(x) \neq 0\}$ is countable. Then there is a bounded linear operator*
$$H_0 : X \longrightarrow c_0(\omega_1)$$
with nonseparable range.

Proof: The assumption means that we can select a sequence $f_\alpha(\alpha < \omega_1)$ of norm-one functionals on X such that:

(34) $(\forall x \in X)(\exists \beta < \omega_1)(\forall \alpha \geq \beta)$ $f_\alpha(x) = 0$.

We plan to use the assumption $\mathfrak{m} > \omega_1$ to select an uncountable subset $\Gamma \subseteq \omega_1$ such that

$$\langle f_\gamma(x) : \gamma \in \Gamma \rangle \in c_0(\Gamma)$$

for all $x \in X$. Thus it is natural to consider the following set $\mathcal{P}$ of finite approximations to such a Γ. The elements of $\mathcal{P}$ are triples

$$p = \langle D_p, \Gamma_p, \epsilon_p \rangle,$$

where D_p is a finite subset of X, where Γ_p is a finite subset of ω_1 and where ϵ_p is a rational number from the interval $(0, 1)$. We order $\mathcal{P}$ by letting $p \leq q$ if:

(35) $D_p \subseteq D_q,\ \ \Gamma_p \subseteq \Gamma_q,\ \ \epsilon_p \geq \epsilon_q.$
(36) $(\forall x \in D_p)(\forall \delta \in \Gamma_q \backslash \Gamma_p)|f_\delta(x)| < \epsilon_p.$

As before, the following claim constitutes an important part of the proof.

Claim 1. $\mathcal{P}$ satisfies the countable claim conditions.

Proof: Let $p_\xi(\xi < \omega_1)$ be a given sequence of elements of $\mathcal{P}$. We need to find $\xi \neq \eta$ such that p_ξ and p_η admit a common extension. Applying a Δ-system argument and going to a subsequence of $p_\xi(\xi < \omega_1)$ we may (and will) assume that there exist $\epsilon \in \mathbb{Q} \cap (0, 1)$, $k \in \mathbb{N}$, and finite sets $D \subseteq X$ and $\Delta \subseteq \omega_1$ such that for all $\xi < \eta < \omega_1$:

(37) $D_{p_\xi} \cap D_{p_\eta} = D,$
(38) $\Gamma_{p_\xi} \cap \Gamma_{p_\eta} = \Delta,$
(39) $|\Gamma_{p_\xi} \backslash \Delta| = k,$
(40) $\Gamma_{p_\xi} \backslash \Delta < \Gamma_{p_\eta} \backslash \Delta.$
(41) $(\forall y \in D_{p_\xi})(\forall \gamma \in \Gamma_{p_\eta} \backslash \Delta)\ f_\gamma(y) = 0.$

For $\xi < \eta < \omega_1$, let

$$p_{\xi\eta} = (D_{p_\xi} \cup D_{p_\eta}, \Gamma_{p_\xi} \cup \Gamma_{p_\eta}, \epsilon).$$

Then $p_{\xi\eta}$ is a member of $\mathcal{P}$ which by (41) extends p_ξ. So it suffices to show that there must exist $\xi < \eta$ such that $p_\eta \leq p_{\xi\eta}$, which is equivalent to

(42) $(\forall x \in D_{p_\eta})(\forall \gamma \in \Gamma_{p_\xi} \backslash \Delta)|f_\gamma(x)| < \epsilon.$

Suppose by way to a contradiction that such a $\xi < \eta$ cannot be found which is equivalent to

(43) $(\forall \xi < \eta)(\exists x \in D_{p_\eta})(\exists \gamma \in \Gamma_{p_\xi} \backslash \Delta)|f_\gamma(x)| \geq \epsilon.$

For an ordinal $\eta < \omega_1$ and a real number $t > 0$, set

$$W_\eta(\geq t) = \{\langle g_0, g_1, \cdots, g_{k-1} \rangle \in (B_{X^*})^k : (\exists x \in D_{p_\eta})(\exists i < k)\ |g_i(x)| \geq t\}$$

Similarly, one defines the strict version $W_\eta(> t)$. Clearly for every $\eta < \omega_1$ and $t > 0$, the set $W_\eta(\geq t)$ is closed and $W_\eta(> t)$ is open in the kth-power $(B_{X^*})^k$ of the dual ball of X with the weak* topology. For $\xi < \omega_1$, let

$$\Gamma_{p_\xi} \backslash \Delta = \{\alpha_0(\xi), \alpha_1(\xi), \cdots, \alpha_{k-1}(\xi)\}$$

be the increasing enumeration according to the natural ordering on the ordinals, and let

$$w_\xi = \langle f_{\alpha_0(\xi)}, f_{\alpha_1(\xi)}, \cdots, f_{\alpha_{k-1}(\xi)} \rangle.$$

Then by (41) and (43), we get the following

(44) $(\forall \xi < \eta)$ $w_\xi \in W_\eta(\geq \epsilon)$,

(45) $(\forall \xi > \eta)$ $w_\xi \notin W_\eta(> \epsilon/2)$.

It follows that for all $\alpha < \omega_1$,

$$\overline{\{w_\xi : \xi \leq \alpha\}} \cap \overline{\{w_\xi : \xi > \alpha\}} = \emptyset,$$

where the closure is taken in $(B_{X^*}, w^*)^k$. So in particular, no complete accumulation point of w_ξ $(\xi < \omega_1)$ in $(B_{X^*})^k$ can belong to any of the sets $\overline{\{w_\xi : \xi \leq \alpha\}}$ for $\alpha < \omega_1$. From this we conclude that the kth-power of B_{X^*}, and therefore, B_{X^*} itself fails to be countably determined, contradicting our initial assumption. □

For $\beta < \omega_1$, set

$$\mathcal{D}_\beta^0 = \{p \in \mathcal{P} : (\exists_\gamma \in \Gamma_p) \gamma \geq \beta\}.$$

Claim 2. $\mathcal{D}_\beta^0$ is a dense subset of $\mathcal{P}$ for all $\beta < \omega_1$.

Proof: Given $p \in \mathcal{P}$ and $\beta < \omega_1$, applying (21) we find $\delta \geq \beta$ such that $f_\delta(x) = 0$ for all $x \in D_p$. Then $q = (D_p, \Gamma_p \cup \{\gamma\}, \epsilon_p)$ is a member of $\mathcal{D}_\beta^0$ which extends p. □

For $x \in X$, set

$$\mathcal{D}_x^1 = \{p \in \mathcal{P} : x \in D_p\}.$$

Claim 3. $\mathcal{D}_x^1$ is a dense subset of $\mathcal{P}$ for all $\beta < \omega_1$.

Proof: Given $p \in \mathcal{P}$ and $x \in X$, let

$$q = (D_p \cup \{x\}, \Gamma_p, \epsilon_p).$$

Then q is a number of $\mathcal{D}_x^1$ extending p. □

For $\epsilon \in \mathbb{Q} \cap (0, 1)$, set

$$\mathcal{D}_\epsilon^2 = \{p \in \mathcal{P} : \epsilon_p \leq \epsilon\}.$$

Claim 4. $\mathcal{D}_\epsilon^2$ is a dense subset of $\mathcal{P}$ for all $\epsilon \in \mathbb{Q} \cap (0,1)$.

Proof: Given $p \in \mathcal{P}$ and $\epsilon \in \mathbb{Q} \cap (0,1)$, let

$$q = (D_p, \Gamma_p, \min\{\epsilon_p, \epsilon\}).$$

Then q is a member of $\mathcal{D}_\epsilon^2$ extending q. $\square$

Let X_0 be a norm-dense subset of X of size dens(X). Since $\omega_1 \le$ dens $(X) <$ m, applying the definition Baire category invariant m, we get a filter $\mathcal{F} \subseteq \mathcal{P}$ such that

(46) $(\forall \beta \in \omega_1)$ $\mathcal{F} \cap \mathcal{D}_\beta^0 \ne \emptyset$,
(47) $(\forall x \in X_0)$ $\mathcal{F} \cap \mathcal{D}_x^1 \ne \emptyset$,
(48) $(\forall \epsilon \in \mathbb{Q} \cap (0,1))$ $\mathcal{F} \cap \mathcal{D}_x^2 \ne \emptyset$.

Let

$$\Gamma = \bigcup_{\Gamma \in \mathcal{F}} \Gamma_p.$$

Then Γ is an uncountable subset with the properly that

$$\langle f_\gamma(x) : \gamma \in \Gamma \rangle \in c_0(\Gamma)$$

for all $x \in X_0$, and therefore for all $x \in X$. The proof of Theorem 72 is finished. $\square$

Now we prove a result which supplements that of Theorem 72 and which however uses the smaller Baire category invariant mm rather than m.

Theorem 73. *Suppose X is a Banach space of density $<$ mm whose dual ball B_{X^*} equipped with the weak* topology is not countably tight. Then there is bounded linear operator*

$$H : X \longrightarrow c_0(\omega_1)$$

with nonseparable range.

Proof: Our topological assumption about (B_{x^*}, w^*) gives us the existence of a non-closed subset $K \subseteq B_{X^*}$ with the property that $K \cap D$ is relatively closed in D for every countable $D \subseteq B_{X^*}$. Note that this is equivalent to saying that K is not weak* closed but it has the property that $\overline{K_0}^{w^*} \subseteq K$ for all countable $K_0 \subseteq K$. Let

$$R = \{f - g : f, g \in K, f \ne g\}.$$

Choose a point $p^* \in \overline{K}^{w^*} \setminus K$. Clearly, we may assume that $p^* = 0^*$ and, increasing K if necessary, we may also assume that K is symmetric in the sense that $-h \in K$ whenever $h \in K$. Let $\mathcal{I}$ be the ideal of all countable subsets A of R which converge to 0^*, i.e., countable subsets A of R with the property that $\{f \in A : |f(x)| \geq \epsilon\}$ is finite for all $x \in X$ and $\epsilon > 0$. By our assumption about the density of our Banach space X, $\mathcal{I}$ is P-ideal being the orthogonal ideal on R generated by $<\mathfrak{m}\mathfrak{m}$ sets, So we are left with the following two alternatives

(a) There is uncountable $S \subseteq R$ such that $[S]^\omega \subseteq \mathcal{I}$, or

(b) R can be decomposed into countably many sets R_n orthogonal to $\mathcal{I}$.

Note that the alternative (a) is giving us the conclusion of Theorem 73 so, it suffices to show that the alternative (b) is false for $\mathcal{I}$. First of all note that since $\mathfrak{p}$ is bigger than the density of X, this alternative is equivalent to the statement that R can be decomposed into countably many sets R_n $(n < \omega)$ such that no countable subset of a fixed R_n accumulated to 0^*. To see this is false, note that we can increase the R_n without changing this basic property and assume that they are symmetric in the sense that $-h \in R_n$ whenever $h \in R_n$. Let us say that subset $Z \subseteq B_{X^*}$ is *countably closed* if $\overline{Z_0}^{w^*} \subseteq Z$ for all countable $Z_0 \subseteq Z$. Note that K is a countably closed (but not closed) subset of B_{X^*}. Choose a maximal filter $\mathcal{F}$ consisting of subsets of K that are countably closed in B_{X^*} such that $\overline{U} \cap K \in \mathcal{F}$ for all weak* open neighborhoods U of 0^*. Note that $\mathcal{F}$ is in fact σ-complete in the sense that the intersection of countably many members of $\mathcal{F}$ belongs to $\mathcal{F}$. Let $\mathcal{F}^+$ be the co-ideal of all subsets Z of K which intersect every member of $\mathcal{F}$. For a subset W of B_{X^*}, let $[W]$ denote the minimal countably closed subset of B_{X^*} containing W, i.e., the union of all closures of countable subsets of W. Note that, by maximality of $\mathcal{F}$, we have that $[W] \in \mathcal{F}$ for all $W \in \mathcal{F}^+$.

Since, $R = \bigcup_{n=0}^\infty R_n$, a simple Fubini-type argument would give us an n, a set $Y \in \mathcal{F}^+$, and a family $Z_f(f \in Y)$ of elements of $\mathcal{F}^+$ such that

(49) $(\forall f \in Y)(\forall g \in Z_f)\ f - g \in R_n$.

Choose countable subsets H of K and a countable subset $\mathcal{F}_0$ of $\mathcal{F}$ such that:

(50) $\mathcal{F}$ is closed under finite intersections and $F \cap H$ is infinite for every $F \in \mathcal{F}$.

(51) $Y \cap H$ is infinite as well as any of the intersections $Z_{y^*} \cap H$ for $y^* \in Y \cap H$.

(52) If W is one of the sets Y or Z_f for $f \in Y \cap H$ then $[W]$ belongs to $\mathcal{F}_0$ and every $h \in [W] \cap H$ belongs to the closure of $W \cap H$.

By our assumption about R_n and the property (49), the countable set

$$A = \{f - g : f \in Y \cap H, g \in Z_f \cap H\}$$

belongs to $\mathcal{I}$. It follows that A does not accumulates to 0^*, so we can find an $\epsilon > 0$ and a finite sequence $x_0, x_1, ..., x_{k-1}$ of elements of our Banach space X such that

(53) $(\forall h \in A)(\exists i < k)\ |h(x_i)| \geq \epsilon.$

By (50), we can chose a point h of K that is a weak*-accumulation point of every set of the form $F \cap H$ for $F \in \mathcal{F}$. Since $[Y] \in \mathcal{F}_0$, we can find $h_0 \in [Y] \cap H$ such that

(54) $(\forall i < k)\ |h_0(x_i) - h(x_i)| < \epsilon/6.$

By (52), we can find $f \in Y \cap H$ such that

(55) $(\forall i < k)\ |f(x_i) - h_0(x_i)| < \epsilon/6.$

It follows that,

(56) $(\forall i < k)\ |f(x_i) - h(x_i)| < \epsilon/3.$

By (52), we know that $[Z_f] \in \mathcal{F}_0$ so we can repeat the argument for this set and get $g \in Z_f$ such that

(57) $(\forall i < k)\ |g(x_i) - h(x_i)| < \epsilon/3.$

Combining (56) and (57), we conclude that

(58) $(\forall i < k)\ |f(x_i) - g(x_i)| < 2\epsilon/3,$

contradicting (53) since clearly the difference $f - g$ belongs to A. This finishes the proof that the alternative (b) is false for $\mathcal{I}$ and R, so we are left with the first. Note that the alternative (a) for the ideal $\mathcal{I}$ on the set R means that there is an uncountable subset W of R such that

(59) $\langle h(x) : h \in W \rangle \in c_0(W)$ for all $x \in X.$

Normalizing and enumerating the elements of W, and using Lemma 1, we get the conclusion of Theorem 73. $\square$

Combining Theorems 71, 72, and 73, we obtain the following result.

Theorem 74. *Every nonseparable Banach space X of density $<$ mm admits a quotient space with a monotone Schauder basis of length ω_1.*

Corollary 75. *If mm $> \omega_1$ then every nonseparable Banach space X contains an uncountable biorthogonal system.*

11.3   A problem of Rolewicz

Let C be a closed convex subset of some Banach space X. A point $x \in C$ is a *support point* of C if there is $f \in X^*$ such that

$$f(x) = \inf\{f(y) : y \in C\} < \sup\{f(y) : y \in C\}.$$

Rolewicz showed that every closed convex subset of a separable Banach space contains a point which is *not* a support point and asked the following question.

Problem (Rolewicz) Does every nonseparable Banach space X contains a closed convex subset composed entirely of support points?

We have the following affirmative answer to this question.

Theorem 76. *If* $\mathrm{mm} > \omega_1$ *then a Banach space X is separable if and only if every closed convex subset C of X contains a point which does not support it.*

Proof: Let X be a nonseparable Banach space. By Corollary 7 we can select a bounded biorthogonal system $(x_\alpha, f_\alpha)(\alpha < \omega_1)$ of X. Let

$$C = \overline{\mathrm{conv}} \{x_\alpha : \alpha < \omega_1\}.$$

We claim that every point x of C is a support point. To see this fix a $\beta < \omega_1$, such that $x \in \overline{\mathrm{conv}} \{x_\alpha : \alpha < \beta\}$. Then

$$0 = f_\beta(x) = \inf\{f_\beta(y) : y \in C\} < 1 = f_\beta(x_\beta) = \sup\{f_\beta(y) : y \in C\}$$

This finishes the proof. $\qquad\qquad\qquad\qquad\qquad\qquad\qquad\qquad\qquad\square$

11.4 Function spaces

The purpose of this section is to examine the biorthogonality problem in the class of Banach spaces of the form $\mathcal{C}(K)$. We start with a result giving a topological condition on K that guarantees the existence of such a system.

Theorem 77. *If a compact space K has a nonseparable subspace then its function space $\mathcal{C}(K)$ contains an uncountable biorthogonal system.*

Proof: We may (and will) assume that K itself is not separable. Note that we may also assume that K satisfies the countable chain condition or else one gets a rather special uncountable biorthogonal system

$$(f_\alpha, \delta_{x_\alpha})(\alpha < \omega_1)$$

in $\mathcal{C}(K)$. These assumptions on K allow us to choose recursively for each countable ordinal α a continuous function $f_\alpha : K \mapsto [0,1]$ and points $x_\alpha, y_\alpha \in K$ such that:

(60) $f_\alpha(x_\alpha) = 1$ and $f_\alpha(y_\alpha) = 0$,
(61) $f_\alpha(x_\beta) = f_\alpha(y_\beta) = 0$ for $\beta < \alpha$,

(62) $f_\alpha(x_\beta) = f_\alpha(y_\beta)$ for $\alpha < \beta$.

Suppose that for some countable ordinal β we have chosen $f_\alpha(\alpha < \beta)$ and $x_\alpha, y_\alpha(\alpha < \beta)$ satisfying these three conditions. The following Claim shows that we can continue the recursive construction.

Claim. *There exist $x \neq y$ in $K \setminus \overline{\{x_\alpha, y_\alpha : \alpha < \beta\}}$ such that $f_\alpha(x) = f_\alpha(y)$ for all $\alpha < \beta$.*

Proof: Otherwise every open subset U of K whose closure is disjoint from the set $\overline{\{x_\alpha, y_\alpha : \alpha < \beta\}}$ is metrizable as its points are separated by the countable family $f_\alpha(\alpha < \beta)$ of continuous functions. So in particular every such open set is separable. Choose a maximal family $\mathcal{U}$ of such open sets which are moreover pairwise disjoint. By our assumption that K satisfies the countable chain condition, the family $\mathcal{U}$ is countable, so the union of $\mathcal{U}$ is separable. It follows that

$$(\bigcup \mathcal{U}) \cup \{x_\alpha, y_\alpha : \alpha < \beta\}$$

is a dense separable subspace of K contradicting our assumption that K is not separable. This finishes the proof of Claim

Having chosen $f_\alpha(\alpha < \omega_1)$ and $x_\alpha, y_\alpha(\alpha < \omega_1)$ satisfying (60), (61) and (62), we form the system

$$(f_\alpha, \delta_{x_\alpha} - \delta_{y_\alpha})(\alpha < \omega_1).$$

Note that this is a biorthogonal system of $\mathcal{C}(K)$.   □

The following result has been already established above using the smaller Baire category invariant $\mathfrak{mm}$.

Theorem 78. *If $\mathfrak{m} > \omega_1$ then a compact space K is metrizable if and only if all biorthogonal systems of the functions space $\mathcal{C}(K)$ are countable.*

Proof: Let K be a given non metrizable compactum. We may assume that K has weight exactly $\aleph_1$ and by Theorem 77 we may (and will) assume that K is hereditarily separable. So, in particular, K and therefore all its finite powers are countably determined. By the argument already appearing above, we can select a sequence $x_\alpha, y_\alpha(\alpha < \omega_1)$ of pairs of different points of K such that

(63) $(\forall f \in \mathcal{C}(K))(\exists \alpha < \omega_1)(\forall \beta \geq \alpha)f(x_\beta) = f(y_\beta)$

For $\alpha < \omega_1$, let

$$\mu_\alpha = \delta_{x_\alpha} - \delta_{y_\alpha}.$$

Let $\mathcal{P}$ be the collection of all triples $p = (D_p, \Gamma_p, \epsilon_p)$ where D_p is a finite subset of $\mathcal{C}(K)$, Γ_p is a finite subset of ω_1, and $\epsilon_p \in \mathbb{Q} \cap (0,1)$. We order $\mathcal{P}$ by letting $p \leq q$ iff $D_p \subseteq D_q$, $\Gamma_p \subseteq \Gamma_q$, $\epsilon_p \geq \epsilon_q$ and

(64) $(\forall f \in D_p)(\forall \alpha \in \Gamma_q \setminus \Gamma_p)|f(x_\alpha) - f(y_\alpha)| < \epsilon_p.$

Claim 1. $\mathcal{P}$ satisfies the countable chain condition.

Proof: Let $p_\xi(\xi < \omega_1)$ be a given sequence of elements of $\mathcal{P}$. Performing a Δ-system argument and going to a subsequence, we may assume to have an integer $k > 0$, rational $\epsilon \in (0,1)$ and finite sets $D \subseteq \mathcal{C}(K)$ and $\Delta \subseteq \omega_1$ such that for all $\xi < \eta < \omega_1$:

(65) $\epsilon_{p_\xi} = \epsilon,$
(66) $\Gamma_{p_\xi} \cap \Gamma_{p_\eta} = \Delta,$
(67) $\Gamma_{p_\xi} \setminus \Delta < \Gamma_{p_\eta} \setminus \Delta,$
(68) $|\Gamma_{p_\xi} \setminus \Delta| = k,$
(69) $D_{p_\xi} \cap D_{p_\xi} = D.$

For $\xi < \eta < \omega_1$, set

$$p_{\xi\eta} = (D_{p_\xi} \cup D_{p_\eta}, \ \Gamma_{p_\xi} \cup \Gamma_{p_\eta}, \epsilon).$$

Note that by (64) and (67), $p_\xi \leq p_{\xi\eta}$ whenever $\xi < \eta$, so we would be done if we could find $\xi < \eta < \omega_1$ such that, moreover, $p_\eta \leq p_{\xi\eta}$. Suppose such $\xi < \eta$ cannot be found. This means that

(70) $(\forall \xi < \eta)(\exists f \in D_{p_\eta})(\exists \alpha \in \Gamma_{p_\eta} \setminus \Delta)|f(x_\alpha) - f(y_\alpha)| \geq \epsilon.$

For $\eta < \omega_1$, set

$$W_\eta = \{\langle a_0, b_0, \cdots, a_{k-1}, b_{k-1}\rangle \in K^{2k} : (\exists f \in D_{p_\eta})(\exists i < k)$$
$$|f(a_i) - f(b_i)| \geq \epsilon\},$$
$$W_\eta^0 = \{\langle a_0, b_0, \cdots, a_{k-1}, b_{k-1}\rangle \in K^{2k} : (\exists f \in D_{p_\eta})(\exists i < k)$$
$$|f(a_i) - f(b_i)| > \epsilon/2\}.$$

Then W_η is a closed subset of K^{2k} and W_η^0 is an open subset of K^{2k} such that $W_\eta \subseteq W_\eta^0$. For $\xi < \omega_1$, let

$$\Gamma_{p_\xi} \setminus \Delta = \{\alpha_0(\xi), \cdots, \alpha_{k-1}(\xi)\}$$

be the increasing enumeration, and let

$$w_\xi = \langle x_{\alpha_0}(\xi), y_{\alpha_0}(\xi), x_{\alpha_1}(\xi), y_{\alpha_1}(\xi), \cdots, x_{\alpha_{k-1}}(\xi), y_{\alpha_{k-1}}(\xi)\rangle.$$

Then (62) and (70) transfer to the following properties

(71) $(\forall \xi < \eta) \ w_\eta \notin W_\xi^0,$
(72) $(\forall \xi < \eta) \ w_\xi \in W_\eta.$

It follows that $w_\xi(\xi < \omega_1)$ is a sequence of elements of K^{2k} such that

$$\overline{\{w_\xi : \xi \leq \alpha\}} \cap \overline{\{w_\xi : \xi > \alpha\}} = 0$$

for all $\alpha < \omega_1$. Hence no complete accumulation point of $w_\xi \ (\xi < \omega_1)$ in K^{2k} can belong to any of the sets $\overline{\{w_\xi : \xi \leq \alpha\}}$ for $\alpha < \omega_1$. It follows that the power K^{2k}, and therefore K itself, fails to be countably determined, a contradiction. This proves Claim 1.

For $\gamma \in \omega_1$, $f \in \mathcal{C}(K)$ and $\epsilon \in \mathbb{Q} \cap (0,1)$, set

$$\mathcal{D}_\gamma^0 = \{p \in \mathcal{P} : (\exists \alpha \in \Gamma_p) \alpha \geq \gamma\},$$

$$\mathcal{D}_f^1 = \{p \in \mathcal{P} : f \in D_p\},$$

$$\mathcal{D}_\epsilon^2 = \{p \in \mathcal{P} : \epsilon_p \geq \epsilon\}.$$

Then as before one proves the following three facts showing that all these sets are dense in $\mathcal{P}$.

Claim 2. $(\forall p \in \mathcal{P})(\forall \gamma \in \omega_1)(\exists q \in \mathcal{D}_\gamma^0)$ $p \leq q$.   □

Claim 3. $(\forall p \in \mathcal{P})(\forall f \in \mathcal{C}(K))(\exists q \in \mathcal{D}_f^1)$ $p \leq q$.   □

Claim 4. $(\forall p \in \mathcal{P})(\forall \epsilon \in \mathbb{Q} \cap (0,1))(\exists q \in \mathcal{D}_\epsilon^2)$ $p \leq q$.   □

Recall that our assumption about K is that it has weight exactly $\aleph_1$ or in other words that the Banach space $\mathcal{C}(K)$ has density $\aleph_1$. So let X_0 be a norm-dense subset of $\mathcal{C}(K)$ of size $\aleph_1$. Applying $\mathfrak{m} > \omega_1$, we get a filter $\mathcal{F}$ of $\mathcal{P}$ such that

(73) $(\forall \gamma < \omega_1)$ $\mathcal{F} \cap \mathcal{D}_\gamma^0 \neq \emptyset$,

(74) $(\forall f \in X_0)$ $\mathcal{F} \cap \mathcal{D}_f^1 \neq \emptyset$,

(75) $(\forall \epsilon \in \mathbb{Q} \cap (0,1))$ $\mathcal{F} \cap \mathcal{D}_\epsilon^2 \neq \emptyset$.

Let $\Gamma = \bigcup_{p \in \mathcal{F}} \Gamma_p$. Then Γ is an uncountable subset of ω_1 with the property that

(76) $\langle f(x_\gamma) - f(y_\gamma) : \gamma \in \Gamma \rangle \in c_0(\Gamma)$ for all $f \in \mathcal{C}(K)$

It follow that the sequence $\mu_\gamma = \delta_{x_\gamma} - \delta_{y_\gamma}(\gamma \in \Gamma)$ of elements of $\mathcal{C}(K)^*$ satisfies the hypothesis of Theorem 71. So by Theorem 71, we obtain a quotient space of $\mathcal{C}(K)$ with a Schauder basis of length ω_1. It follows that $\mathcal{C}(K)$ has an uncountable biorthogonal system.   □

Note that if in the above proof we have started with a 0-dimensional compactum K then already on the basis of (76), we could have obtained an uncountable bi-orthogonal system in $\mathcal{C}(K)$, i.e., relying on Theorem 71 becomes unnecessary. To see this pick for each $\gamma \in \Gamma$ a clopen set V_γ containing x_γ but not y_γ. Applying (76) to the characteristic functions of the sets $V_\gamma(\gamma \in \Gamma)$ and going to an uncountable subset of Γ_0 of Γ we get

(77) $(\forall \gamma \neq \delta \in \Gamma_0)(x_\gamma \in V_\delta \longleftrightarrow y_\gamma \in V_\delta)$.

So, in particular, $(x_{V_\gamma}, \delta_{x_\gamma} - \delta_{y_\gamma})(\gamma \in \Gamma_0)$ forms an uncountable biorthogonal system in $\mathcal{C}(K)$. Note also that if we let $\mathcal{B}$ be the corresponding Boolean algebra of clopen subsets of K, the sequence $V_\gamma(\gamma \in \Gamma)$ of elements of $\mathcal{B}$ has the property that no V_γ is in the subalgebra of $\mathcal{B}$ generated by $V_\delta(\delta \notin \gamma)$. Thus, we have the following result.

Theorem 79. *If* $\mathfrak{m} > \omega_1$ *then every uncountable Boolean algebra* $\mathcal{B}$ *contains an uncountable subset* $\mathcal{I}$ *such that no* $a \in \mathcal{I}$ *belongs to a subalgebra generated by* $\mathcal{I} \setminus \{a\}$.

Chapter 12

Structure of Compact Spaces

12.1 Covergence in topology

The proof of the following result is a variation of the S-space proof, adjusted to the context where some sort of compactness is used.

Theorem 80. *If* $\mathfrak{mm} > \omega_1$, *then every countably compact subspace* Y *of a compact countably tight space* X *is Lindelöf and therefore compact.*

Proof. Suppose that Y is not Lindelöf and pick a family $\mathcal{U}$ of open subsets of X covering Y such that $Y \setminus \bigcup \mathcal{U}_0 \neq \emptyset$, for every countable $\mathcal{U}_0 \subseteq \mathcal{U}$. Choose also a well-ordering $<_w$ of X.

Choose a maximal filter $\mathcal{F}$ of closed subsets of X such that:

(a) $\overline{F \cap Y} = F$, for all $F \in \mathcal{F}$.

(b) $\overline{Y \setminus \bigcup \mathcal{U}_0} \in \mathcal{F}$, for all countable $\mathcal{U}_0 \subseteq \mathcal{U}$.

As in the case of the S-space proof, for each $y \in Y$ pick open (in X) neighborhoods U_y, V_y such that:

(c) $y \in U_y \subseteq \overline{U_y} \subseteq V_y$.

(d) $\overline{Y \setminus V_y} \in \mathcal{F}$.

Let $\mathcal{P}$ be the collection pof all pairs $p = (H_p, \mathcal{N}_p)$ such that H_p is a finite subset of Y and $\mathcal{N}_p$ is a finite ϵ-chain of countable elemtary submodels of some fixed large-enough structure (H_θ, ϵ) containing all the relevant objects, such that

(e) H_p is separated by $\mathcal{N}_p$.

(f) If $x \in H_p$ and if $N = max\{M \in \mathcal{N}_p : x \notin M\}$ then x is the $<_w$-minimal point of the intersection

$$Y \cap \bigcap \{\overline{F \cap N} : F \in \mathcal{F} \cap N\}.$$

For $p \in \mathcal{P}$ and $N \in \mathcal{N}_p$, we use the notation x_N^p for the $<_w$-minimal point Y belonging to the intersection of the family $\{\overline{F \cap N} : F \in \mathcal{F} \cap N\}$, and also use the notation,

$$W_N^p = \bigcap\{U_x : x \in H_p \setminus N \ \& \ x_N^p \in U_x\}.$$

For $p, q \in \mathcal{P}$, set $q \leq p$ iff $H_q \supseteq H_p$, $\mathcal{N}_q \supseteq \mathcal{N}_p$ and if the following holds for every $N \in \mathcal{N}_p$,

(g) $(H_q \setminus H_p) \cap N \subseteq W_N^p \cup \bigcup\{U_z : z \in H_p \cap N\}.$

Note that $\leq$ is a transitive relation on $\mathcal{P}$, so we have a partial ordering that is subject to our Baire category assumption provided we show the following.

Claim 1. $\mathcal{P}$ is stationary-set preserving.

Proof. Let $C(p)$ $(p \in \mathcal{P})$ be a given $\mathcal{P}$-club. Let $\bar{p} \in \mathcal{P}$ and $E \in Stat(\omega_1)$. Choose a countable elementary submodel $M \prec (H_{(2^\theta)^+}, \epsilon)$ containing all these objects such that $\delta = M \cap \omega_1 \in E$.

Find $q \leq (H_{\bar{p}}, \mathcal{N}_{\bar{p}} \cup \{M \cap H_\theta\})$ such that $C(q) \setminus \delta \neq \emptyset$. We claim that $\delta \in C(q)$ and this will finish the proof. Suppose not and let $r \leq q$ and $\beta < \delta$ be such that

$$C(s) \cap [\beta, \delta) = \emptyset \text{ for all } s \leq r.$$

Let $n = |H_r \setminus M|$. Note that $n \geq 1$ since otherwise the elementarity of M would give us that $C(p)$ $(p \in \mathcal{P})$ fails the unboundedness property. Let

$$\mathcal{J} = \{K \in [Y]^{\leq n} : (\exists s \leq_{end} r \cap M) \ [C(s) \setminus \beta \neq \emptyset \ \& \ K \subseteq H_s \setminus H_{r \cap M}\}.$$

Note that $\mathcal{J} \in M \cap H_\theta$. So we need to find a $K \in \mathcal{J} \cap [Y]^n \cap M$ such that $K \subseteq W_M^r.$[1]

As before we apply the Fubini argument and trim $\mathcal{J}$ to a subset $\mathcal{J}_0$ such that

(h) $\emptyset \in \mathcal{J}_0$.

(i) $\overline{\{y \in Y : K \cup \{y\} \in \mathcal{J}_0\}} \in \mathcal{F}$ for all $K \in \mathcal{J}_0$ of size $< n$.

For $K \in \mathcal{J}_0$, set

$$Y_K = \{y \in Y : K \cup \{y\} \in \mathcal{J}_0\}.$$

Note that $Y_\theta \in M \cap H_\theta$ and since $M \cap H_\theta \in \mathcal{N}_r$ and $Y_\theta \in \overline{\mathcal{F} \cap M \cap H_\theta}$ by (f) we conclude that $x_M^r \in \overline{Y_\theta \cap M}$. Since W_M^r is a neighborhood of x_M^r, we can find $z \in Y_\theta \cap M$ belonging to W_M^r. Applying countable tightness of X while working in M we can find countable $Z_0 \subseteq Y_\theta \cap M$ such that $z \in \overline{Z_0}$. So there is $y_0 \in Z_0 \cap W_M^r$. Since $y_0 \in M$, the set $Y_{\{y_0\}}$ also belongs to M so we can continue the process. After n steps we arrive at $\{y_0, \ldots, y_{n-1}\}$ in $\mathcal{J}_0 \cap [Y]^n \cap M$ such that $\{y_0, \ldots, y_{n-1}\} \subseteq W_M^r$. So we can find $\bar{r} \in M \cap \mathcal{P}$ end-extending $r \cap M$ such that $C(\bar{r}) \setminus \beta \neq \emptyset$ and $H_{\bar{r}} \setminus H_{r \cap M} = \{y_0, \ldots, y_{n-1}\}$. Then

$$s = (H_{\bar{r}} \cup H_r, \mathcal{N}_{\bar{r}} \cup \mathcal{N}_r)$$

is an element of $\mathcal{P}$ extending both $\bar{r}$ and r such that

$$C(s) \cap [\beta, \delta) \supseteq C(\bar{r}) \cap [\beta, \delta) \neq \emptyset,$$[2]

a contradiction. This completes the proof. $\qquad \square$

For $\alpha < \omega_1$, let

$$\mathcal{D}_\alpha = \{p \in \mathcal{P} : \exists N \in \mathcal{N}_p \ (\alpha \in N \ \& \ H_p \setminus N \neq \emptyset)\}.$$

Claim 2. Each $\mathcal{D}_\alpha$ is dense in $\mathcal{P}$.

---

[1]Here, of course, $W_M^r = W_{M \cap H_\theta}^r$.
[2]This follows from the elementarity of M since $\bar{r} \in M$, $C(\bar{r}) \neq \emptyset$ and $\delta = M \cap \omega_1$.

Proof. Consider $p \in \mathcal{P}$. Find a countable elementary submodel $N \prec H_\theta$ containing α and p. By countable compactness of Y, the intersection

$$\bigcap \{\overline{F \cap N} : F \in \mathcal{F} \cap N\}$$

must have a point in Y so we take the $<_w$-minimal one and call it x_N. Then

$$q = (H_p \cup \{x_N\}, \mathcal{N}_p \cup \{N\})$$

is an extension of p belonging to $\mathcal{D}_\alpha$. $\square$

Since $\mathfrak{mm} > \omega_1$ there is a filter $\mathcal{G} \subseteq \mathcal{P}$ such that $\mathcal{G} \cap \mathcal{D}_\alpha \neq \emptyset$. Let

$$H_\mathcal{G} = \bigcup_{p \in \mathcal{G}} H_p \text{ and } \mathcal{N}_\mathcal{G} = \bigcup_{p \in \mathcal{G}} \mathcal{N}_p.$$

Note that $\mathcal{N}_\mathcal{G}$ is an ϵ-chain of countable elementary submodels of H_θ, so we can enumerate it as $(N_\alpha^\mathcal{G} : \alpha \in \omega_1)$ according to ϵ. The set $H_\mathcal{G}$ is also uncountable so we can enumerate it according to the separation by $\mathcal{N}_\mathcal{G}$ as $\langle x_\alpha^\mathcal{G} : \alpha \in \omega_1 \rangle$. The following claim will give us the desired contradiction since the compactum X is assumed to be countably tight.

Claim 3. $\langle x_{\alpha+1}^\mathcal{G} : \alpha \in \omega_1 \rangle$ is a free sequence.

Proof. Suppose there is $\alpha \in \omega_1$ such that

$$\{x_\xi^\mathcal{G} : \xi < \alpha\} \cap \{x_\xi^\mathcal{G} : \xi > \alpha\} \neq \emptyset.$$

We assume that α is minimal such. Choose $p \in \mathcal{G}$ such that $x_\alpha^\mathcal{G} \in H_p$ and $N = max\{M \in \mathcal{N}_p : x_\alpha^\mathcal{G} \notin M\}$. Then $x_\alpha^\mathcal{G} \in W_N^p$ and $\overline{W_N^p} \subseteq V_{x_\alpha^\mathcal{G}}$. Note that by (d) and (f), $x_\xi^\mathcal{G} \notin V_{x_\alpha^\mathcal{G}}$ for all $\xi > \alpha$, so in particular $x_\xi^\mathcal{G} \notin \overline{W_N^p}$ for all $\xi > \alpha$. On the other hand, by (g), every $x_\xi^\mathcal{G}$ for $\xi < \alpha$ either belongs to W_N^p or to H^p or to some U_z for $z \in H_p \cap N$. Since H_p is finite and since $x_\xi^\mathcal{G} \in X \setminus V_z$ for all $\xi < \alpha$ and $z \in H_p \cap N$, the $x_\xi^\mathcal{G}$ ($\xi < \alpha$) that belong to $\bigcup_{z \in H_p \cap N} U_z$ do not accumulate to a point of $\overline{\{x_\xi^\mathcal{G} : \xi > \alpha\}}$. It follows that $\{x_\xi^\mathcal{G} : \xi < \alpha\} \cap W_N^p$ accumulates to a point of $\overline{\{x_\xi^\mathcal{G} : \xi > \alpha\}}$, a contradiction. $\square$

This finishes the proof.

                    $\square$

Corollary. *If* $\mathfrak{mm} > \omega_1$ *then every compact countably tight space is sequential.*

Proof. Let X be a given compact countably tight space, and let $Y \subseteq X$ be a given sequentially closed subset. By the previous theorem it suffices to show that Y is countably compact. Suppose Y is not countably compact and choose a countable set $D \subseteq Y$ with no accumulation point in Y. Choose a countably compact subspace W of $\overline{D}$ containing D such that $|W| \leq c$. By the previous theorem, W is compact and therefore closed in X. It follows that $W = \overline{D}$. Since $\overline{D} \setminus D$ is closed and has size $c = \omega_2$ it must have a point y of relative character $\leq \aleph_1$. Since $\overline{D} \setminus D$ is a G_δ subset of $\overline{D}$, the character of y in $\overline{D}$ is at most $\aleph_1$. By $\mathfrak{mm} > \omega_1$ there is a sequence $\{x_n\}_n^\infty \subseteq D$ converging to y. Since Y is sequentially closed we conclude that $y \in Y$, contadicting the assumption that D has no accumulation points in Y. $\square$

Theorem 81. *If $\mathfrak{mm} > \omega_1$ then every compact ccc T_5 space is Fréchet.*

Proof. By previous results, every compact ccc T_5 space X is countably tight and therefore separable and sequential. To see that X is Fréchet consider $D \subseteq X$ and $y \in \overline{D} \setminus D$. Since $\overline{D} \setminus \{y\}$ is not closed, there is a sequence $\{x_n\}_n^\infty \subseteq \overline{D} \setminus \{y\}$ converging to y. Since $X \setminus \{y\}$ is normal and $\{x_n : n \in \omega\}$ is discrete in $X \setminus \{y\}$ there is a sequence $\{U_n\}_n^\infty$ of open subsets of $X \setminus \{y\}$ such that $x_n \in U_n$ for all n and such that $\{U_n\}_n^\infty$ is a discrete collection in $X \setminus \{y\}$. Then for each n, we can choose $y_n \in U_n \cap D$. Then it follows that $y_n \to y$ as required. $\square$

12.2 Ultrapowers versus reduced powers

For a nonprincipal filter $\mathcal{F}$ on ω, let $\omega^\omega / \mathcal{F}$ denote the corresponding reduced power, the quotient of ω^ω relative to the equivalence relation

$$f =_\mathcal{F} g \text{ iff } \{n : f(n) = g(n)\} \in \mathcal{F}$$

and the ordering defined by

$$[f]_\mathcal{F} < [g]_\mathcal{F} \ ^3 \text{ iff } \{n : f(n) < g(n)\} \in \mathcal{F}.$$

As customary in this area we shall prefer to work with ω^ω rather than $\{[g]_\mathcal{F} : g \in \omega^\omega\}$ and for $f, g \in \omega^\omega$ write $f <_\mathcal{F} g$ whenever $\{n : f(n) < g(n)\} \in \mathcal{F}$. As it is also customary in this area, when $\mathcal{F}$ is the filter of cofinite subsets of ω we write $<_{FIN}$ or more often $<^*$ for the corresponding ordering of eventual dominance of ω^ω.

---

[3] $[g]_\mathcal{F} = \{h \in \omega^\omega : h =_\mathcal{F} g\}$.

Theorem 82. *If* $\mathrm{mm} > \omega_1$ *then* $\omega^\omega / \mathcal{U} \not\hookrightarrow \omega^\omega / FIN$, *for every nonprincipal ultrafilter* $\mathcal{U}$ *on* ω.

The proof is based on the following two lemmas.

Theorem 83. *If* $\mathrm{mm} > \omega_1$ *then* $(2^{\omega_1}, <_{lex}) \hookrightarrow (\omega^\omega / \mathcal{U}, <_\mathcal{U})$, *for every nonprincipal ultrafilter* $\mathcal{U}$ *on* ω.

Theorem 84. *If* $\mathrm{mm} > \omega_1$ *then* $(2^{\omega_1}, <_{lex}) \not\hookrightarrow (\omega^\omega, <_{FIN})$.

Proof of Theorem 83. Recursively on the nodes of the complete binary tree $(2^{\omega_1}, \sqsubseteq)$, we choose a sequence

$$(f_t, g_t) \in (\omega^\omega)^2 \quad (t \in 2^{\omega_1})$$

such that

(1) $s \sqsubseteq t$ implies $f_s <^* f_t <^* g_t <^* g_s$.

(2) $f_t <^* f_{t^\smallfrown 0} <^* g_{t^\smallfrown 0} <^* f_{t^\smallfrown 1} <^* g_{t^\smallfrown 1} <^* g_t$.

(3) $f(n) \le g(n)$, for all n.

If for each $x \in 2^{\omega_1}$ we can find $h_x \in \omega^\omega$ such that

$$f_{x \restriction \alpha} < h_x < g_{x \restriction \alpha}$$

then $x \to h_x$ will be an embedding of $(2^\omega, <_{lex})$ into $(\omega^\omega / \mathcal{U}, <_\mathcal{U})$. So it suffices to show the following fact

Sublemma. Suppose $\mathrm{m} > \omega_1$ and that $(f_\alpha, g_\alpha) \in (\omega^\omega)^2$ $(\alpha < \omega_1)$ is a sequence satisfying

(a) $f_\alpha(n) \le g_\alpha(n)$ for all n.

(b) $f_\alpha <^* f_\beta <^* g_\beta <^* g_\alpha$, for all $\alpha < \beta$.

Then for every nonprincipal ultrafilter $\mathcal{U}$ on ω there is $h \in \omega^\omega$ such that

$$f_\alpha <_\mathcal{U} h <_\mathcal{U} g_\alpha, \quad \text{for all } \alpha < \omega_1.$$

Proof. Let $\mathcal{P}$ be the collection of all finite subsets p of ω_1 such that for all $\{\alpha, \beta, \gamma\} \in [p]^3$ and all $n \in \omega$, two of the intervals from

$$\{[f_\alpha(n), g_\alpha(n)], \ [f_\beta(n), g_\beta(n)], \ [f_\gamma(n), g_\gamma(n)]\}$$

have nonempty intersection. We claim that $\mathcal{P}$ satisfies the countable chain condition. Consider a sequence p_ξ $(\xi < \omega_1)$ of elements of $\mathcal{P}$. Refining the sequence, we may assume that the p_ξ's are all of the same size k and that for some $m \in \omega$ and $\{(s_i, t_i) : i < k\} \in (\omega^m \times \omega^m)^k$ and all $\xi < \omega_1$:

(c) If α is the i-th element of p_ξ then $s_i \subseteq f_\alpha$ and $t_i \subseteq g_\alpha$.

(d) $f_\alpha(n) \leq f_\beta(n) \leq g_\beta(n) \leq g_\alpha(n)$, for all $\alpha, \beta \in p$ and $n \geq m$.

It follows then that $p_\xi \cup p_\eta \in \mathcal{P}$ for all ξ and η. By our assumption $\mathfrak{m} > \omega_1$ there is an uncountable $\Gamma \subseteq \omega_1$ such that $\{\{\alpha\} : \alpha \in \Gamma\}$ is a cetered subset of $\mathcal{P}$. This means that for all $\{\alpha, \beta, \gamma\} \in [\Gamma]^3$ and all $n < \omega$ two of the intervals from the set

$$\{[f_\xi(n), g_\xi(n)] : \xi \in \{\alpha, \beta, \gamma\}\}$$

intersect. Fix $n \in \omega$. Consider the family

$$\mathcal{J}_n = \{[f_\alpha(n), g_\alpha(n)] : \alpha \in \Gamma\}$$

of intervals. Let I_n be the anti-lexicographically minimal interval of $\mathcal{J}_n$ and let

$$\mathcal{J}_n^0 = \{I \in \mathcal{J}_n : I \cap I_n \neq \emptyset\} \text{ and}$$

$$\mathcal{J}_n^1 = \{I \in \mathcal{J}_n : I \cap I_n = \emptyset\}.$$

By the property of Γ it follows easily that both of these two families have the 2-intersection property. So by Helly's theorem

$$h_i(n) = min(\bigcap \mathcal{J}_n^i) \quad (i = 0, 1)$$

exists $(min \ \emptyset = 0)$. Thus we have just defined two functions $h_0, h_1 \in \omega^\omega$ such that for every $\alpha \in \Gamma$ and every n either

$$f_\alpha(n) \leq h_0(n) \leq g_\alpha(n), \text{ or}$$

$$f_\alpha(n) \leq h_1(n) \leq g_\alpha(n).$$

So if $\mathcal{U}$ is a nonprincipal ultrafilter on ω there will be $i \in \{0, 1\}$ and uncountable $\Gamma_0 \subseteq \Gamma$ such that $f_\alpha(n) \leq h_i(n) \leq g_\alpha(n)$ for all $\alpha \in \Gamma_0$. $\quad\square$

This finishes the proof of Theorem 83. $\quad\square$

Proof of Theorem 84. Suppose that there is an embedding

$$\Phi : (2^{\omega_1}, <_{lex}) \to (\omega^\omega, <^*)$$

and work towards a contradiction.

For $t \in 2^{<\omega_1}$ we choose $f_t, g_t \in \omega^\omega$ such that

(1) $f_t(n) \leq g_t(n)$ for all n.

(2) $\Phi(t\hat{}0) =^* f_t$ and $\Phi(t\hat{}1) =^* g_t$.

Let $\mathcal{P}$ be the collection of all pairs $p = (H_p, \mathcal{N}_p)$ where:

(3) H_p is a finite chain of the complete binary tree $2^{<\omega_1}$,

(4) $\mathcal{N}_p$ is a finite ϵ-chain of countable elementary submodels of $(H_{(2^{\aleph_1})^+}, \epsilon)$.

(5) $\mathcal{N}_p$ separates H_p.

(6) $(\forall N \in \mathcal{N}_p)(\forall t \in H_p \setminus N)(\forall \mathcal{D} \in N \cap DO(2^{<\omega_1}))$ [4] $(\exists \alpha < N \cap \omega_1)\, t \upharpoonright \alpha \in \mathcal{D}$.

We order $\mathcal{P}$ by letting $p \leq q$ iff

(7) $H_p \supseteq H_q$, $\mathcal{N}_p \supseteq \mathcal{N}_q$.

(8) $(\forall t \in H_q)(\forall s \in H_p \setminus H_q)\ [s \subseteq t \to (\exists k)\, f_s(k) > g_t(k)]$.

Claim 4. $\mathcal{P}$ is stationary-set preserving.

Proof. Let $C(p)\ (p \in \mathcal{P})$ be a given $\mathcal{P}$-club, $\bar{p} \in \mathcal{P}$ be a given condition and $S \subseteq \omega_1$ a given stationary set. Pick a countable elementary submodel $M \prec (H_{(2^{2^{\aleph_1}})^+}, \epsilon)$ such that $\delta = M \cap \omega_1 \in S$ and such that M contains all the objects accumulated so far. Choose $t_M \in 2^\delta$ such that for all $\mathcal{D} \in DO(2^{<\omega_1} \cap M)$ there is $\alpha < \delta$ such that $t_M \upharpoonright \alpha \in \mathcal{D}$. Then

$$(H_{\bar{p}} \cup \{t_M\}, \mathcal{N}_{\bar{p}} \cup \{M \cap H_{(2^{\aleph_1})^+}\})$$

is a member of $\mathcal{P}$ so we can find its extension q such that $C(q) \setminus \delta \neq \emptyset$. We claim that $\delta \in C(q)$. Otherwise, we can find $r \leq q$ and $\beta < \delta$ such that

(9) $C(r') \cap [\beta, \delta] = \emptyset$ for all $r' \leq r$.

Let $n = |H_r \setminus M|$. Choose $m < \omega$ such that

(10) $(\forall k \geq m)(\forall u, v \in H_r)\ [u \subseteq v \to f_u(k) \leq g_v(k)]$.

Let $\mathcal{X}$ be the collection of all n-element chains F of the tree $2^{<\omega_1}$ satisfying (10) in place of H_r for which we can find an end-extension $\bar{r}$ of $r \cap M$ such that $C(\bar{r}) \setminus \beta \neq \emptyset$ and $F = H_{\bar{r}} \setminus H_{r\cap M}$. Clearly, $\mathcal{X} \in M \cap H_{(2^{\aleph_1})^+}$ and $H_r \setminus M \in \mathcal{X}$. We would get the desired contradiction if we can find an $F \in \mathcal{X} \cap M$ such that

(11) $(\forall u \in F)(\forall v \in H_r \setminus M)\ u \sqsubseteq v$,

---
[4] $DO(2^{<\omega_1})$ denotes the family of all dense-open subsets of the complete binary tree $2^{<\omega_1}$.

(12) $u = min(F) \;\&\; v = max(H_r \setminus M) \;\rightarrow\; (\exists k \geq m)\; f_u(k) > g_v(k)$.

If such an $F \in \mathcal{X} \cap M$ can be found, choose a witness $\bar{r} \in \mathcal{P} \cap M$ for its existence. Then

$$r' = (H_{\bar{r}} \cup H_r, \mathcal{N}_{\bar{r}} \cup \mathcal{N}_r)$$

is a condition that extends both $\bar{r}$ and r and therefore has the property

$$C(r') \cap [\beta, \delta] \supseteq C(\bar{r}) \cap [\beta, \delta] \neq \emptyset,$$

contradicting the choice of r. So from now on we assume that there is no $F \in \mathcal{X} \cap M$ satisfying (11) and (12) and work for a contradiction. Let

$$\mathcal{X}_0 = \{F \in \mathcal{X} : (\forall E \in \mathcal{X})(max(E) < min(F) \rightarrow (\forall k \geq m)\; f_{min(E)}(k)$$
$$\leq g_{max(F)}(k))\}.$$

Then $\mathcal{X}_0 \in M \cap H_{(2^{\aleph_1})^+}$ and $H_r \setminus M \in \mathcal{X}_0$. For $F \in \mathcal{X}$, let $h_F \in \omega^\omega$ be defined by letting $h_F(k) = 0$ for $k < m$ and

$$h_F(k) = sup\{f_{min(E)}(k) : E \in \mathcal{X}_0 \;\&\; max(E) \sqsubseteq min(F)\}.$$

Note that

(13) $h_F(k) \leq g_{max(F)}(k)$ for all k.

Let

$$\mathcal{X}_1 = \{F \in \mathcal{X}_0 : (\forall k \geq m)\; h_F(k) \geq f_{min(E)}(k)\}.$$

Then by our assumption that there is no $F \in \mathcal{X} \cap M$ satisfying (11) and (12) and the choice of $t_M (= min(H_r \setminus M))$ we conclude that $H_r \setminus M \in \mathcal{X}_1$. Using again the genericity of t_M (i.e., that $(\forall \mathcal{D} \in DO(2^{<\omega_1}))(\exists \alpha < \delta)\; t_M \upharpoonright \alpha \in \mathcal{D}$) we get the following

(14) $(\forall E, F \in \mathcal{X}_1)\; [t_M \sqsubseteq min(E), min(F) \;\rightarrow\; h_E = h_F]$.

Using the genericity of t_M and the fact that $\mathcal{X}_1 \in M$, we can find $t_0 \sqsubset t$ such that

$$\mathcal{D} = \{min(F) : F \in \mathcal{X}_1\}$$

is dense above t_0. So in particular we can find $E, F \in \mathcal{X}_1$ such that

(15) $t\hat{\ }0 \sqsubseteq min(E)$ and $t\hat{\ }1 \sqsubseteq min(F)$.

Find $m_1 \geq m$ such that for all $k \geq m_1$,

(16) $f_{t^\frown 0}(k) \leq f_{min(E)}(k) \leq h_E(k) \leq g_{max(E)}(k) \leq g_{t^\frown 0}(k)$.

(17) $g_{t^\frown 0}(k) \leq f_{t^\frown 1}(k)$,

(18) $f_{t^\frown 1}(k) \leq f_{min(F)}(k) \leq h_F(k) \leq g_{max(F)}(k) \leq g_{t^\frown 1}(k)$.

Note that this contradicts (14). This finishes the proof. $\qquad\qquad$ □

For $\alpha < \omega_1$, set

$$\mathcal{D}_\alpha = \{p \in \mathcal{P} : (\exists N \in \mathcal{N}_p)(\alpha \in N \ \& \ H_p \setminus N \neq \emptyset)\}.$$

Then each $\mathcal{D}_\alpha$ is a dense-open subset of $\mathcal{P}$. Using $\mathfrak{mm} > \omega_1$ we can find a filter $\mathcal{G} \subseteq P$ such that $\mathcal{G} \cap \mathcal{D}_\alpha \neq \emptyset$ for all $\alpha < \omega_1$. Let

$$C = \bigcup_{p \in \mathcal{G}} H_p.$$

Then C is an uncountable chain of $2^{<\omega_1}$ such that

(19) $(\forall u \sqsubseteq v \in C)(\exists k) \ f_u(k) > g_v(k)$.

Let $X = \bigcup C \in 2^{\omega_1}$ and let $h = \Phi(X) \ (\in \omega^\omega)$. Then by the choice of Φ and $(f_t, g_t) \ (t \in 2^{<\omega_1})$ we conclude that

(20) $(\forall u \sqsubseteq \in C) \ f_u <^* h <^* g_u$.

Thus we can find uncountable $C_0 \subseteq C$, an integer m, and $\sigma, \tau \in \omega^\omega$ such that

(21) $(\forall k \geq m)(\forall u \in C_0) \ f_u(k) \leq h(k) \leq g_u(k)$.

(22) $(\forall u \in C_0) \ (f_u \upharpoonright m = \sigma \ \& \ g_u \upharpoonright m = \tau)$.

Note that by (1) and (22) we in particular have that $\sigma(k) \leq \tau(k)$ for all k. Combining this with (21) we get

(23) $(\forall u \sqsubseteq v \in C)(\forall k \in \omega) \ f_u(k) \leq g_v(k)$.

contradicting (20). This finishes the proof of Theorem 84. $\qquad\qquad$ □

12.3 Automatic continuity in Banach algebras

Recall the notion of Banach algebra $(\mathcal{B}, +, \cdot, \|\cdot\|)$, a Banach space with the additional operator of multiplication and the norm $\|\cdot\|$ which has the property that

$$\|x \cdot y\| \leq \|x\| \cdot \|y\|$$

for all $x, y \in \mathcal{B}$. Typical example is the space $\mathcal{C}(K)$ of all continuous functions on K with the operation of pointwise multiplication and the supremum norm $\|\cdot\|_\infty$. In this section we use the Baire category assumption $\mathfrak{m} \mathfrak{m} > \omega_1$ to study the following classical question.

Question: (Kaplansky 1948) Is every norm on a Banach algebra of the form $\mathcal{C}(K)$ equivalent[5] to the usual supremum norm on $\mathcal{C}(K)$?

Theorem 85. *If* $\mathfrak{m} \mathfrak{m} > \omega_1$ *then every norm on a Banach algebra of the form* $\mathcal{C}(K)$ *is equivalent to the supremum norm.*

Proof. We shall use the criterion of Bade, Curtis and Johnson which says that if there is a nonequivalent norm on some $\mathcal{C}(K)$ then there is a non-principal ultrafilter $\mathcal{U}$ on ω, a radical[6] Banach algebra $\mathcal{B}$ and a nontrivial homomorphism

$$\Phi : c_0/\mathcal{U} \to \mathcal{B}.$$

Take $a = (a_n) \in c_0^+$ such that $\Phi(a/\mathcal{U}) \neq \emptyset$. Pick a strictly increasing sequence (n_k) of positive integers such that

(1) $(\forall k)(\forall n \geq n_k) \, a_n < (1/k)^{k^2}$

Let $g \in \omega^\omega$ be such that $g^{-1}(k) = [n_k, n_{k+1})$. Define

$$\Psi : \{f \in \omega^\omega : f < g\} \to c_0^+$$

by letting $\Psi(f) = (a_n^{f(n)/g(n)^2})$. (Here $f < g$ denotes the fact $f(n) < g(n)$ for all n such that $g(n) \neq 0$.)

Claim 5. $f_0 <_\mathcal{U} f_1 \ \to \ \Psi(f_0)/\mathcal{U} \mid \Psi(f_1)/\mathcal{U}$.

Proof. To see this define $z = (z_n) \in c_0$ by

$$z_n = a_n^{(f_1(n) - f_0(n))/g(n)^2}$$

and note that $\Psi(f_0) \cdot z/\mathcal{U} = \Psi(f_1)/\mathcal{U}$. $\quad\square$

---

[5]Recall that two norms $\|\cdot\|$ and $\|\|\cdot\|\|$ are equivalent if we can find constants $c_0, c_1 > 0$ such that $c_0\|x\| \leq \|\|x\|\| \leq c_1\|x\|$ for all x.

[6]$\mathcal{B}$ is **radical** if $\|a_n\|^{1/n} \to 0$ for all $a \in \mathcal{B}$.

Claim 6. For no $f < g$ the image $\Phi(\Psi(f)/\mathcal{U}$ is a nilpotent[7] element of $\mathcal{B}$.

Proof. We shall prove that, in fact

(2) $\Psi(f)^m \mid a$ for all m (in c_0).

Fix m. Let $z = (z_n) \in \mathbb{R}^\omega$ be defined by

$$z_n = a_n^{(1-mf(n))/g(n)^2}.$$

Then

(3) $z_n \le a_n^{1-\frac{n}{g(n)}} \le a_n^{1/2}$ for all but finitely many n.

Since $a_n^{1/2} \to 0$ we conclude that $z \in c_0^+$. Since $\Psi(f)^m \cdot z = a$, the proof is finished.   □

Let

$$\mathcal{A} = \{b \in \mathcal{B} : (\forall n)\ b^n \ne 0\}.$$

Then we have shown that

(4) $\Phi(\Psi(f)/\mathcal{U}) \in \mathcal{A}$ for all $f < g$.

and that $\Phi \circ \Psi$ leads us to a strictly increasing embedding

$$\{f : f < g\}/\mathcal{U} \hookrightarrow (\mathcal{A}, |).$$

This would contradict the theorem of the previous section once we show the following.

Claim 7. There is a strictly increasing $\Sigma : (\mathcal{A}, |) \to (\omega^\omega, <^*)$.

Proof. Set $\Sigma(a)(n) = \lceil 1/\|a^n\| \rceil$.   □

This finishes the proof.   □

Corollary. *If* $\mathfrak{m}\mathfrak{m} > \omega_1$ *then every homomorphism* Φ *from a Banach algebra* $\mathcal{C}(K)$ *into a Banach algebra* $\mathcal{B}$ *is continuous.*

---

[7]Recall that $a \in \mathcal{B}$ is **nilpotent** if $a^n = 0$ for some n.

Proof. Given a homomorphism $\Phi : \mathcal{C}(K) \to \mathcal{B}$ define a new norm $\||| \cdot \|||$ on $\mathcal{C}(K)$ by

$$\||| f \||| = max\{\|f\|_\infty, \|\Phi(f)\|_\mathcal{B}\}.$$

By the previous result, $\||| \cdot \|||$ is equivalent to the norm $\| \cdot \|_\infty$ of $\mathcal{C}(K)$, i.e., for some constants $c_0, c_1 > 0$ and all $f \in \mathcal{C}(K)$ we have $c_0\|f\|_\infty \le \||| f \||| \le c_1\|f\|_\infty$. It follows that for every $f \in \mathcal{C}(K)$,

$$\|\Phi(f)\|_\mathcal{B} \le \||| f \||| \le c_1\|f\|_\infty.$$

Hence Φ is continuous.

$\square$

Chapter 13

Ramsey Theory on Ordinals

13.1   The arrow notation

The Ramsey theory of ordinal numbers is often known as Partition Calculus. The reader is referred to [35] and [6] for a good introduction to the Ramsey theory on ordinals. For α, β and γ ordinals the *arrow notation*

$$\alpha \to (\beta, \gamma)^2$$

expresses the statement that for every coloring $[\alpha]^2 = K_0 \cup K_1$, either there is a set $B \subseteq \alpha$ of order type β such that $[B]^2 \subseteq K_0$ or there is a set $C \subseteq \alpha$ of order type γ such that $[C]^2 \subseteq K_1$. Given ordinals α and β, let

$$(\beta, \gamma)^2 = \min\{\alpha : \alpha \to (\beta, \gamma)^2\}.$$

These are so-called *Ramsey numbers*. Some examples of finite Ramsey numbers are:

$$(3, 3)^2 = 6$$
$$(3, 4)^2 = 9$$
$$(4, 4)^2 = 18.$$

It is interesting that the Ramsey number $(5, 5)^2$ is still unknown. Note also that we have in general

$$(m, n)^2 \leq (m, n - 1)^2 + (m - 1, n)^2.$$

13.2 $\omega_2 \nrightarrow (\omega_2, \omega + 2)^2$

The main theorems of this and the next section will combine to show that under the assumption $\mathfrak{mm} > \omega_1$, there is a substantial combinatorial difference between ω_1 and ω_2. We will show that $\mathfrak{mm} > \omega_1$ implies the following two statements:

$$\omega_1 \rightarrow (\omega_1, \alpha)^2 \text{ for all } \alpha < \omega_1$$
$$\omega_2 \nrightarrow (\omega_2, \omega + 2)^2.$$

Notation. We will use the symbol $\subseteq_{ctbl}$ for "subset mod countable", so that for any sets A and B, we define

$$A \subseteq_{ctbl} B \iff |A \setminus B| \leq \aleph_0$$
$$\iff A \subseteq B \cup H \text{ for some countable set } H.$$

Theorem 86.[1] *If* $\mathfrak{mm} > \omega_1$ *then*

$$\omega_2 \nrightarrow (\omega_2, \omega + 2)^2.$$

Proof. In Theorem 101 below, we shall see that our assumption $\mathfrak{mm} > \omega_1$ implies $2^{\aleph_0} = \aleph_2$. So the collection $[\omega_2]^{\leq \aleph_1}$ has cardinality $\aleph_2$, and so we can let $\{G_\alpha\}_{\alpha < \omega_2}$ be an enumeration of $[\omega_2]^{\leq \aleph_1}$ such that $G_\alpha \subseteq \alpha$ for all α.

We will recursively construct sequences $\{F_\alpha\}_{\alpha < \omega_2}$ and $\{\Sigma_\alpha\}_{\alpha < \omega_2}$ of subsets of ω_2 satisfying the following conditions:

1. $F_\alpha \subseteq \alpha$ for all α, and

2. $|F_\alpha \cap F_\beta| < \aleph_0$ for all $\alpha \neq \beta$, and

3. $F_\alpha \cap G_\xi \neq \emptyset$ for all α and all $\xi \in \Sigma_\alpha$, and

4. $\Sigma_\alpha \subseteq \alpha$ for all α, and

5. for all β and all $\xi < \beta$, we have

$$\xi \in \Sigma_\beta \iff (\forall A \in [\beta]^{\aleph_0}) \left[G_\xi \not\subseteq_{ctbl} \bigcup_{\alpha \in A} F_\alpha \right],$$

meaning in words that ξ is in Σ_β iff G_ξ cannot be covered mod countable by countably many of the sets F_α with $\alpha < \beta$.

---

[1]Recall that $\omega_2 \rightarrow (\omega_2, \omega + 1)^2$ - see [6]; so this theorem negates the next logical extension of this result.

Fix $\beta < \omega_2$, and suppose that we have chosen sequences $\{F_\alpha\}_{\alpha < \beta}$ and $\{\Sigma_\alpha\}_{\alpha < \beta}$ satisfying the above conditions. We must construct F_β and Σ_β satisfying the required conditions, using a Baire category argument based on the assumption $\mathfrak{m} > \omega_1$.

First, we can easily define the set

$$\Sigma_\beta = \left\{ \xi < \beta : (\forall A \in [\beta]^{\aleph_0}) \left[G_\xi \not\subseteq_{ctbl} \bigcup_{\alpha \in A} F_\alpha \right] \right\},$$

and it is clear that conditions (4) and (5) are satisfied for Σ_β. It remains to construct F_β so that the first 3 conditions are satisfied.

Claim 86.1. *We can construct a set X such that:*

1. $X \subseteq \beta$, and

2. $|F_\alpha \cap X| \leq \aleph_0$ for all $\alpha < \beta$, and

3. $|G_\xi \cap X| = \aleph_1$ for all $\xi \in \Sigma_\beta$.

Proof. Since $\beta < \omega_2$, we have $|\beta| \leq \omega_1$, and so we can enumerate

$$\beta = \{\alpha_\eta\}_{\eta < \omega_1}.$$

Since $\Sigma_\beta \subseteq \beta$, we also have $|\Sigma_\beta| \leq |\beta| \leq \omega_1$, and so we can enumerate

$$\Sigma_\beta = \{\xi_\eta\}_{\eta < \omega_1},$$

and we can insist that this enumeration be done in such a way that each $\xi \in \Sigma_\beta$ appears as ξ_η for uncountably many η.

We will now recursively define the set

$$X = \{x_\eta\}_{\eta < \omega_1} \subseteq \beta.$$

Fix $\eta < \omega_1$, and suppose that we have chosen elements $x_\gamma \in \beta$ for $\gamma < \eta$. We must show how to choose x_η.

Since $\eta < \omega_1$, we have $|\eta| \leq \aleph_0$, and so

$$\{\alpha_\gamma : \gamma < \eta\} \in [\beta]^{\leq \aleph_0}.$$

Since $\xi_\eta \in \Sigma_\beta$, it follows from the definition of Σ_β that

$$G_{\xi_\eta} \not\subseteq_{ctbl} \bigcup_{\gamma < \eta} F_{\alpha_\gamma}.$$

We also have

$$|\{x_\gamma : \gamma < \eta\}| \leq \aleph_0,$$

and so we can conclude

$$G_{\xi_\eta} \not\subseteq \bigcup_{\gamma < \eta} F_{\alpha_\gamma} \cup \{x_\gamma : \gamma < \eta\}.$$

We therefore choose x_η to be the least ordinal of

$$G_{\xi_\eta} \setminus \left(\bigcup_{\gamma < \eta} F_{\alpha_\gamma} \cup \{x_\gamma : \gamma < \eta\} \right).$$

We must now show that X satisfies the required conditions:

1. For each η, we have $x_\eta \in G_{\xi_\eta} \subseteq \xi_\eta \subseteq \beta$.

2. Suppose $\alpha < \beta$. We have $\alpha = \alpha_\eta$ for some $\eta < \omega_1$. For any $\gamma > \eta$, we have by choice of x_γ that $x_\gamma \notin F_{\alpha_\eta} = F_\alpha$. So we have

   $$F_\alpha \cap X = F_\alpha \cap \{x_\gamma : \gamma \leq \eta\},$$

   but we know $\eta < \omega_1$ and so $|\eta| \leq \aleph_0$, and so it follows that $|F_\alpha \cap X| \leq \aleph_0$.

3. Suppose $\xi \in \Sigma_\beta$. We have $\xi = \xi_\eta$ for uncountably many η. For each such η we have by choice of x_η that $x_\eta \in G_{\xi_\eta} = G_\xi$, and also that x_η is distinct from x_γ for any $\gamma < \eta$. It follows that $G_\xi \cap X$ is uncountable, and since $|G_\xi| \leq \aleph_1$, we have $|G_\xi \cap X| = \aleph_1$.

So X is as required.   □

Let $\mathcal{P}$ be the collection of all pairs $p = \langle A_p, \Delta_p, \rangle$ where

1. A_p is a finite subset of X, and

2. Δ_p is a finite subset of β.

We order $\mathcal{P}$ by defining $p \leq q$ iff

1. $A_p \supseteq A_q$, and

2. $\Delta_p \supseteq \Delta_q$, and

3. $(A_p \setminus A_q) \cap \bigcup_{\alpha \in \Delta_q} F_\alpha = \emptyset$.

Claim 86.2. *The relation $\leq$ is a partial-order relation on $\mathcal{P}$.*

Proof. Reflexivity is trivial.

   Antisymmetry follows easily from the first two conditions.

To show transitivity: Suppose $p \leq q \leq r$. We must show that $p \leq r$. The first two conditions are trivial. To show that the last condition holds, notice that since $\Delta_r \subseteq \Delta_q$, we have

$$(A_p \setminus A_r) \cap \bigcup_{\alpha \in \Delta_r} F_\alpha \subseteq \left((A_p \setminus A_q) \cup (A_q \setminus A_r) \right) \cap \bigcup_{\alpha \in \Delta_r} F_\alpha$$

$$= \left((A_p \setminus A_q) \cap \bigcup_{\alpha \in \Delta_r} F_\alpha \right) \cup \left((A_q \setminus A_r) \cap \bigcup_{\alpha \in \Delta_r} F_\alpha \right)$$

$$\subseteq \left((A_p \setminus A_q) \cap \bigcup_{\alpha \in \Delta_q} F_\alpha \right) \cup \left((A_q \setminus A_r) \cap \bigcup_{\alpha \in \Delta_r} F_\alpha \right)$$

$$= \emptyset \cup \emptyset = \emptyset$$

and so we conclude that $p \leq r$ and so $\leq$ is transitive. $\square$

Claim 86.3. *The partial order $\langle \mathcal{P}, \leq \rangle$ satisfies the countable chain condition.*

Proof. We will show that any uncountable subset of $\mathcal{P}$ contains two compatible elements and is therefore not an antichain. Suppose that $Y \subseteq \mathcal{P}$ is uncountable.

First, for each $p \in Y$, consider the integer $|A_p|$. There can be only countably many distinct integers, but Y is uncountable, so by the Pigeonhole Principle we can fix an uncountable subset $Y_0 \subseteq Y$ and some integer k such that $|A_p| = k$ for all $p \in Y_0$.

Now, consider the set

$$\left\{ (A_p \times \{0\}) \cup (\Delta_p \times \{1\}) : p \in Y_0 \right\}.$$

This set is an uncountable collection of finite sets, and so by the Δ-System Lemma it must include an uncountable subset forming a Δ-system. This means that there is an uncountable subset $Y_1 \subseteq Y_0$ and there are sets A and Δ, such that for all $p \neq q \in Y_1$ we have

$$A_p \cap A_q = A \text{ and } \Delta_p \cap \Delta_q = \Delta.$$

Now we will recursively define a subset

$$Y_2 = \{p_\xi\}_{\xi < \omega_1} \subseteq Y_1$$

as follows: Fix $\xi < \omega_1$, and suppose that we have chosen elements $p_\eta \in Y_1$ for all $\eta < \xi$. We must show how to choose p_ξ.

Since $\xi < \omega_1$, we have $|\xi| \leq \aleph_0$. For each $\eta < \xi$, we have Δ_{p_η} is finite. For each $\alpha \in \Delta_{p_\eta} \subseteq \beta$ we know by definition of X that $|F_\alpha \cap X| \leq \aleph_0$. So we know

$$\left| \bigcup_{\eta < \xi} \bigcup_{\alpha \in \Delta_{p_\eta}} F_\alpha \cap X \right| \leq \aleph_0.$$

For each $p \in Y_1$, we know A_p is a finite subset of X, and the sets $(A_p \setminus A)$ are pairwise disjoint, and so of the uncountably many $p \in Y_1 \setminus \{p_\eta : \eta < \xi\}$ there must be some p such that

$$(A_p \setminus A) \cap \left(\bigcup_{\eta < \xi} \bigcup_{\alpha \in \Delta_{p_\eta}} F_\alpha \cap X \right) = \emptyset.$$

Let p_ξ be this p. This completes the recursion, and so we have constructed our uncountable subset $Y_2 \subseteq Y_1$.

For any $\xi < \omega_1$, we have by construction of p_ξ that

$$(A_{p_\xi} \setminus A) \cap \left(\bigcup_{\eta < \xi} \bigcup_{\alpha \in \Delta_{p_\eta}} F_\alpha \cap X \right) = \emptyset,$$

and since $A_{p_\xi} \subseteq X$ it follows that

$$(A_{p_\xi} \setminus A) \cap \bigcup_{\eta < \xi} \bigcup_{\alpha \in \Delta_{p_\eta}} F_\alpha = \emptyset.$$

We must now show that we can find $\eta < \xi \in \omega_1$ such that p_η and p_ξ are compatible. First, for any $\eta < \xi \in \omega_1$ we can define

$$r_{\eta,\xi} = \left\langle A_{p_\eta} \cup A_{p_\xi}, \Delta_{p_\eta} \cup \Delta_{p_\xi} \right\rangle \in \mathcal{P}.$$

If we can find $\eta < \xi$ such that $r_{\eta,\xi} \leq p_\eta, p_\xi$ then we are done.

Notice that for any $\eta < \xi \in \omega_1$, we have

$$\begin{aligned} A_{r_{\eta,\xi}} \setminus A_{p_\eta} &= \left(A_{p_\eta} \cup A_{p_\xi} \right) \setminus A_{p_\eta} \\ &= A_{p_\xi} \setminus A_{p_\eta} \\ &= A_{p_\xi} \setminus A. \end{aligned}$$

Similarly, we have

$$\begin{aligned} A_{r_{\eta,\xi}} \setminus A_{p_\xi} &= \left(A_{p_\eta} \cup A_{p_\xi} \right) \setminus A_{p_\xi} \\ &= A_{p_\eta} \setminus A_{p_\xi} \\ &= A_{p_\eta} \setminus A. \end{aligned}$$

As shown above, by construction of p_ξ, we know that

$$\left(A_{p_\xi} \setminus A\right) \cap \bigcup_{\alpha \in \Delta_{p_\eta}} F_\alpha = \emptyset,$$

and therefore we have

$$\left(A_{r_{\eta,\xi}} \setminus A_{p_\eta}\right) \cap \bigcup_{\alpha \in \Delta_{p_\eta}} F_\alpha = \emptyset,$$

and it follows that $r_{\eta,\xi} \leq p_\eta$. So all we need to do is show that there are $\eta < \xi \in \omega_1$ such that $r_{\eta,\xi} \leq p_\xi$.

Suppose there do not exist $\eta < \xi \in \omega_1$ such that $r_{\eta,\xi} \leq p_\xi$, and we will derive a contradiction. This means that for every $\eta < \xi \in \omega_1$ we have $r_{\eta,\xi} \not\leq p_\xi$, so that

$$\left(A_{r_{\eta,\xi}} \setminus A_{p_\xi}\right) \cap \bigcup_{\alpha \in \Delta_{p_\xi}} F_\alpha \neq \emptyset.$$

As shown before, this is the same as

$$\left(A_{p_\eta} \setminus A\right) \cap \bigcup_{\alpha \in \Delta_{p_\xi}} F_\alpha \neq \emptyset.$$

With the additional assumption that $\eta \geq 1$, we have, by construction of p_η,

$$\left(A_{p_\eta} \setminus A\right) \cap \bigcup_{\alpha \in \Delta_{p_0}} F_\alpha = \emptyset,$$

and since $\Delta \subseteq \Delta_{p_0}$, it follows that

$$\left(A_{p_\eta} \setminus A\right) \cap \bigcup_{\alpha \in \Delta} F_\alpha = \emptyset.$$

Combining these two results, for $1 \leq \eta < \xi < \omega_1$, we can choose some $\alpha_{\eta,\xi} \in \Delta_{p_\xi} \setminus \Delta$ such that

$$\left(A_{p_\eta} \setminus A\right) \cap F_{\alpha_{\eta,\xi}} \neq \emptyset.$$

We set $|A| = j$, and then, for all $\xi < \omega_1$, since we have $\left|A_{p_\xi}\right| = k$ we can write

$$A_{p_\xi} \setminus A = \{a_{\xi,j}, a_{\xi,j+1}, \ldots, a_{\xi,k-1}\}.$$

[Note that $a_{\xi,i} \neq a_{\xi',i'}$ for any $(\xi, i) \neq (\xi', i')$.] So for any $1 \leq \eta < \xi \in \omega_1$ we can choose some $i_{\eta,\xi} \in \{j, \ldots, k-1\}$ such that

$$a_{\eta,i_{\eta,\xi}} \in F_{\alpha_{\eta,\xi}}.$$

Choose $\mathcal{U} \in \beta\omega \setminus \omega$. In words, let $\mathcal{U}$ be a non-principal ultrafilter on ω. (A non-principal ultrafilter on a set W is an ultrafilter on W that is not of the form $\{Z \subseteq W : w \in Z\}$ for some fixed $w \in W$. We know there exists a non-principal ultrafilter on any infinite set as follows: Start by taking the filter of co-finite subsets, and then using Zorn's Lemma extend it to a maximal proper filter.)

Now, fix $\xi \geq \omega$. For each $i \in \{j, \ldots, k-1\}$ and each $\alpha \in \Delta_{p_\xi} \setminus \Delta$, define

$$B_{\xi,i,\alpha} = \{\eta \in \omega \setminus \{0\} : i_{\eta,\xi} = i \text{ and } \alpha_{\eta,\xi} = \alpha\}$$
$$= \{\eta \in \omega \setminus \{0\} : a_{\eta,i} \in F_\alpha\}.$$

We have thus partitioned $\omega \setminus \{0\}$ into finitely many pairwise disjoint subsets. It follows that we can choose some $\overline{i_\xi} < k$ and $\overline{\alpha_\xi} \in \Delta_{p_\xi} \setminus \Delta$ such that

$$B_{\xi,\overline{i_\xi},\overline{\alpha_\xi}} \in \mathcal{U}.$$

(This is because if every one of the subsets were not in the ultrafilter then all of their complements would have to be in the ultrafilter, and there are only finitely many of these subsets, so their intersection must be in the ultrafilter. But the intersection of all the complements of the partition subsets of $\omega \setminus \{0\}$ is the finite set $\{0\}$, and a finite set cannot a member of a non-principal ultrafilter.)

There are only finitely many possible values for $\overline{i_\xi} < k$ but there are uncountably many $\xi \geq \omega$. By the pigeon hole principle there must be a particular $\overline{i} < k$ and an uncountable subset $K \subseteq \omega_1 \setminus \omega$ such that $\overline{i} = \overline{i_\xi}$ for all $\xi \in K$. In particular, we can find $\xi_0 \neq \xi_1 \in K$ such that $\overline{i_{\xi_0}} = \overline{i} = \overline{i_{\xi_1}}$. This means

$$B_{\xi_0,\overline{i},\overline{\alpha_{\xi_0}}} \in \mathcal{U} \text{ and } B_{\xi_1,\overline{i},\overline{\alpha_{\xi_1}}} \in \mathcal{U},$$

and since the ultrafilter must be closed under finite intersections, we have

$$B_{\xi_0,\overline{i},\overline{\alpha_{\xi_0}}} \cap B_{\xi_1,\overline{i},\overline{\alpha_{\xi_1}}} \in \mathcal{U}.$$

This means

$$\left\{\eta \in \omega \setminus \{0\} : a_{\eta,\overline{i}} \in F_{\overline{\alpha_{\xi_0}}} \cap F_{\overline{\alpha_{\xi_1}}}\right\} \in \mathcal{U}.$$

But $\mathcal{U}$ is a non-principal ultrafilter, so it cannot contain any finite subset of ω. So there are infinitely many η such that

$$a_{\eta,\overline{i}} \in F_{\overline{\alpha_{\xi_0}}} \cap F_{\overline{\alpha_{\xi_1}}},$$

and so it follows that $F_{\overline{\alpha_{\xi_0}}} \cap F_{\overline{\alpha_{\xi_1}}}$ is infinite. We have $\overline{\alpha_{\xi_0}} \in \Delta_{p_{\xi_0}} \setminus \Delta$ and $\overline{\alpha_{\xi_1}} \in \Delta_{p_{\xi_1}} \setminus \Delta$. Since $\xi_0 \neq \xi_1$, it follows that $\overline{\alpha_{\xi_0}} \neq \overline{\alpha_{\xi_1}}$. But this contradicts condition (2) from the construction of the sequence $\{F_\alpha\}$.

We have derived a contradiction, and so we have shown that the uncountable set Y must contain two compatible conditions. This means that $\mathcal{P}$ satisfies the countable chain condition. $\square$

We will now define our dense-open sets in $\langle \mathcal{P}, \leq \rangle$:
For each $\alpha < \beta$, define the set

$$\mathcal{D}_\alpha = \{p \in \mathcal{P} : \alpha \in \Delta_p\}.$$

For each $\xi \in \Sigma_\beta$, define the set

$$\mathcal{E}_\xi = \{p \in \mathcal{P} : A_p \cap G_\xi \neq \emptyset\}.$$

Claim 86.4. *For each $\alpha < \beta$, the set $\mathcal{D}_\alpha$ is dense-open in $\langle \mathcal{P}, \leq \rangle$.*

Proof. To show that $\mathcal{D}_\alpha$ is dense: Suppose $q \in \mathcal{P}$. We must find $p \leq q$ such that $p \in \mathcal{D}_\alpha$. Simply define

$$p = \langle A_q, \Delta_q \cup \{\alpha\} \rangle.$$

It is clear that $p \leq q$ and that $p \in \mathcal{D}_\alpha$. So we have shown that $\mathcal{D}_\alpha$ is dense.

To show that $\mathcal{D}_\alpha$ is open: Suppose that $p \leq q \in \mathcal{P}$ and $q \in \mathcal{D}_\alpha$. It is clear that $\alpha \in \Delta_q \subseteq \Delta_p$, and so $p \in \mathcal{D}_\alpha$. So $\mathcal{D}_\alpha$ is open. $\square$

Claim 86.5. *For each $\xi \in \Sigma_\beta$, the set $\mathcal{E}_\xi$ is dense-open in $\langle \mathcal{P}, \leq \rangle$.*

Proof. To show that $\mathcal{E}_\xi$ is dense: Suppose $q \in \mathcal{P}$. We must find $p \leq q$ such that $p \in \mathcal{E}_\xi$.

By definition of X, since $\xi \in \Sigma_\beta$ we have $|G_\xi \cap X| = \aleph_1$. We also have $|F_\alpha \cap X| \leq \aleph_0$, for all $\alpha < \beta$, in particular for $\alpha \in \Delta_q$. So we clearly have

$$\left| \bigcup_{\alpha \in \Delta_q} (F_\alpha \cap X) \right| \leq \aleph_0,$$

so we can choose some

$$\eta \in (G_\xi \cap X) \setminus \left(\bigcup_{\alpha \in \Delta_q} (F_\alpha \cap X) \right),$$

and it follows that

$$\eta \in (G_\xi \cap X) \setminus \left(\bigcup_{\alpha \in \Delta_q} F_\alpha \right).$$

We now define

$$p = \langle A_q \cup \{\eta\}, \Delta_q \rangle.$$

Since $\eta \in X$, we have $A_p = A_q \cup \{\eta\} \subseteq X$, and so $p \in \mathcal{P}$. We have $\eta \in A_p \cap G_\xi$, so clearly $A_p \cap G_\xi \neq \emptyset$ and so $p \in \mathcal{E}_\xi$. Since $\eta \notin \bigcup_{\alpha \in \Delta_q} F_\alpha$, we have

$$(A_p \setminus A_q) \cap \bigcup_{\alpha \in \Delta_q} F_\alpha = \emptyset,$$

and so it follows that $p \leq q$, and we have shown that $\mathcal{E}_\xi$ is dense.

To show that $\mathcal{E}_\xi$ is open: Suppose that $p \leq q \in \mathcal{P}$ and $q \in \mathcal{E}_\xi$. It is clear that $A_q \cap G_\xi \neq \emptyset$, and since $A_p \supseteq A_q$, it follows that $A_p \cap G_\xi \neq \emptyset$, and so $p \in \mathcal{D}_\xi$. So $\mathcal{E}_\xi$ is open. $\qquad\square$

We know $|\beta| \leq \aleph_1$ and $|\Sigma_\beta| \leq \aleph_1$, so our collection

$$\{\mathcal{D}_\alpha : \alpha < \beta\} \cup \{\mathcal{E}_\xi : \xi \in \Sigma_\beta\}$$

consists of at most $\aleph_1$ dense sets. We have shown that our partial order $\langle \mathcal{P}, \leq \rangle$ satisfies the countable chain condition, and we have assumed that $\omega_1 < \mathfrak{m}$, so there must be a filter $\mathcal{G} \subseteq \mathcal{P}$ that intersects all of our dense sets, that is $\mathcal{G} \cap \mathcal{D}_\alpha \neq \emptyset$ for all $\alpha < \beta$, and $\mathcal{G} \cap \mathcal{E}_\xi \neq \emptyset$ for all $\xi \in \Sigma_\beta$.

Let

$$F_\beta = \bigcup_{p \in \mathcal{G}} A_p.$$

Claim 86.6. *The set F_β satisfies all the required conditions.*

Proof. We will check the conditions in order:

1. For each p we have $A_p \subseteq X \subseteq \beta$, so clearly $F_\beta \subseteq \beta$.

2. Suppose $\alpha < \beta$. Since $\mathcal{G} \cap \mathcal{D}_\alpha \neq \emptyset$, we can choose $p \in \mathcal{G}$ with $\alpha \in \Delta_p$. We will show that $F_\alpha \cap F_\beta \subseteq A_p$.

   Suppose $\eta \in F_\alpha \cap F_\beta$. Since $\eta \in F_\beta$, we have $\eta \in A_q$ for some $q \in \mathcal{G}$. Since $p, q \in \mathcal{G}$ and $\mathcal{G}$ is a filter, there must be an $r \in \mathcal{G}$ with $r \leq p, q$. Since $r \leq p$ and $\alpha \in \Delta_p$, we have

   $$(A_r \setminus A_p) \cap F_\alpha = \emptyset.$$

   Since $\eta \in A_q$ and $r \leq q$, we have $\eta \in A_r$, and so $\eta \in A_r \cap F_\alpha$. It must then be that $\eta \in A_p$.

   We have shown that $F_\alpha \cap F_\beta \subseteq A_p$, and since A_p is finite, it follows that $F_\alpha \cap F_\beta$ is finite.

3. Suppose $\xi \in \Sigma_\beta$. Since $\mathcal{G} \cap \mathcal{E}_\xi \neq \emptyset$, we can choose $p \in \mathcal{G}$ with $A_p \cap G_\xi \neq \emptyset$. But then $A_p \subseteq F_\beta$, so clearly $F_\beta \cap G_\xi \neq \emptyset$.

So F_β satisfies all the required conditions and our recursive construction is complete. $\qquad\square$

It follows that we can construct sequences $\{F_\alpha\}_{\alpha < \omega_2}$ and $\{\Sigma_\alpha\}_{\alpha < \omega_2}$ of subsets of ω_2 satisfying the required conditions (1) through (5).

In order to prove our theorem, we must construct a colouring on pairs from ω_2, and then show that the colouring has no homogeneous sets of the required sizes.

Define a colouring
$$c : [\omega_2]^2 \to \{0, 1\}$$
by setting, for $\alpha < \beta \in \omega_2$,
$$c(\{\alpha, \beta\}) = 1 \iff \alpha \in F_\beta.$$

We must show that this colouring has neither a 0-homogeneous set of size ω_2 nor a 1-homogeneous set of order type $\omega + 2$.

Claim 86.7. *For any set $Y \subseteq \omega_2$ of size ω_2, we can find $\beta \in Y$ and $\xi \in \Sigma_\beta$ such that $G_\xi \subseteq Y$.*

Proof. We will recursively construct sequences
$$\{\xi_\gamma\}_{\gamma < \omega_1} \subseteq \omega_2 \text{ and } \{A_\gamma\}_{\gamma < \omega_1} \subseteq [\omega_2]^{\leq \aleph_0},$$
where the sets A_γ are pairwise disjoint, until we obtain our desired result.

Fix $\gamma < \omega_1$, and suppose we have chosen sequences $\{\xi_\eta\}_{\eta < \gamma}$ and $\{A_\eta\}_{\eta < \gamma}$. We will construct ξ_γ and A_γ.

We know that γ is a countable ordinal, and for each $\eta < \gamma$, we know $A_\eta \in [\omega_2]^{\leq \aleph_0}$. It follows that
$$\bigcup_{\eta < \gamma} A_\eta \in [\omega_2]^{\leq \aleph_0},$$

as it is a countable union of countable sets. We can then define
$$s = \sup \left(\bigcup_{\eta < \gamma} A_\eta \right) + 1 < \omega_2,$$

so that $A_\eta \subseteq s$ for all $\eta < \gamma$. Since $|s| \leq \omega_1$ and $|Y| = \omega_2$, we have $|Y \setminus s| = \omega_2$, and we can let G be the initial segment of $Y \setminus s$ of order type ω_1. Since $G \in [\omega_2]^{\omega_1}$, we can choose some $\xi_\gamma < \omega_2$ such that $G = G_{\xi_\gamma}$.

Clearly we have $G_{\xi_\gamma} \subseteq Y$, and so if we can find $\beta \in Y$ such that $\xi_\gamma \in \Sigma_\beta$, then our Claim would be proven. In that case, we end the recursive construction now because we are done. (Computer scientists do not like breaking out of recursive constructions this way.)

Otherwise, since $|Y| = \omega_2$, we can choose $\beta \in Y$ such that $\beta > \xi_\gamma$, and then we have $\xi_\gamma \notin \Sigma_\beta$ because we already ruled out the other possibility. By definition of Σ_β, it follows that we can choose some $A \in [\beta]^{\aleph_0} \subseteq [\omega_2]^{\aleph_0}$ such that
$$G_{\xi_\gamma} \subseteq_{ctbl} \bigcup_{\alpha \in A} F_\alpha.$$

Now, for any $\alpha < s$ we have $F_\alpha \subseteq \alpha \subseteq s$. But $G_{\xi_\gamma} \cap s = \emptyset$ so it follows that $G_{\xi_\gamma} \cap F_\alpha = \emptyset$. We can let
$$A_\gamma = A \setminus s \in [\omega_2]^{\leq \aleph_0},$$

and we then have

$$G_{\xi_\gamma} \subseteq_{ctbl} \bigcup_{\alpha \in A_\gamma} F_\alpha.$$

For any $\eta < \gamma$, since $A_\eta \subseteq s$, it follows that $A_\eta \cap A_\gamma = \emptyset$. This completes the recursion step.

So we have constructed sequences $\{\xi_\gamma\}_{\gamma<\omega_1} \subseteq \omega_2$ and $\{A_\gamma\}_{\gamma<\omega_1} \subseteq [\omega_2]^{\leq \aleph_0}$, with the latter sequence consisting of pairwise disjoint subsets. We now define

$$G = \bigcup_{\gamma<\omega_1} G_{\xi_\gamma}.$$

Once again we have $G \in [\omega_2]^{\omega_1}$, and so we can choose some $\overline{\xi} < \omega_2$ such that $G = G_{\overline{\xi}}$.

Clearly we have $G_{\overline{\xi}} \subseteq Y$. We then choose $\beta \in Y$ such that $\beta > \overline{\xi}$. We will now show that $\overline{\xi} \in \Sigma_\beta$.

Suppose $\overline{\xi} \notin \Sigma_\beta$ and we will derive a contradiction. By definition of Σ_β, we can choose some $\overline{A} \in [\beta]^{\aleph_0}$ such that

$$G_{\overline{\xi}} \subseteq_{ctbl} \bigcup_{\alpha \in \overline{A}} F_\alpha.$$

The set $\overline{A}$ is countable, and the collection $\{A_\gamma : \gamma < \omega_1\}$ is an uncountable collection of pairwise disjoint sets. It follows that we can find some $\gamma < \omega_1$ such that $\overline{A} \cap A_\gamma = \emptyset$. By the earlier construction of A_γ, we know that

$$G_{\xi_\gamma} \subseteq_{ctbl} \bigcup_{\alpha \in A_\gamma} F_\alpha.$$

Since we know $|G_{\xi_\gamma}| = \omega_1$, and we also know that A_γ is countable, it follows by the Pigeonhole Principle that there must be some $\overline{\alpha} \in A_\gamma$ such that $|G_{\xi_\gamma} \cap F_{\overline{\alpha}}| = \aleph_1$. We then have

$$G_{\xi_\gamma} \cap F_{\overline{\alpha}} \subseteq G_{\xi_\gamma} \subseteq G_{\overline{\xi}} \subseteq_{ctbl} \bigcup_{\alpha \in \overline{A}} F_\alpha.$$

Since we know that $\overline{A}$ is countable, it follows again by the Pigeonhole Principle that there must be some $\alpha \in \overline{A}$ such that

$$|G_{\xi_\gamma} \cap F_{\overline{\alpha}} \cap F_\alpha| = \aleph_1.$$

Certainly then $F_{\overline{\alpha}} \cap F_\alpha$ is uncountable. But we have $\overline{\alpha} \in A_\gamma$ and $\alpha \in \overline{A}$, so clearly $\overline{\alpha} \neq \alpha$. This contradicts condition (2) in the construction of the sequence $\{F_\alpha\}$.

So we have derived a contradiction, so it follows that $\overline{\xi} \in \Sigma_\beta$, and our claim follows. $\square$

Claim 86.8. *The colouring c has no 0-homogeneous set of size ω_2.*

Proof. Suppose $Y \subseteq \omega_2$ is set of size ω_2. We will show that Y cannot be 0-homogeneous. From the previous claim, we can fix $\beta \in Y$ and $\xi \in \Sigma_\beta$ such that $G_\xi \subseteq Y$. Since $\xi \in \Sigma_\beta$, we have by condition (3) of the construction that $F_\beta \cap G_\xi \neq \emptyset$. Since $G_\xi \subseteq Y$, it follows that $F_\beta \cap Y \neq \emptyset$. So we can choose $\alpha \in F_\beta \cap Y$. So we have $\alpha, \beta \in Y$, and since $\alpha \in F_\beta$, it follows by definition of c that $c(\{\alpha, \beta\}) = 1$. This shows that Y is not 0-homogeneous. $\square$

Claim 86.9. *The colouring c has no 1-homogeneous set of order type $\omega + 2$.*

Proof. Suppose $Z \subseteq \omega_2$ is a 1-homogeneous set of order type $\omega + 2$ and we will derive a contradiction.

Since Z has order type $\omega + 2$, we can enumerate

$$Z = \{z_\xi\}_{\xi < \omega + 2},$$

where the enumeration preserves the order of Z as a subset of ω_2. Since Z is 1-homogeneous, we have in particular that for each $n < \omega$,

$$c(\{z_n, z_\omega\}) = 1 = c(\{z_n, z_{\omega+1}\}).$$

Since our enumeration of Z preserves the order, we know that $z_n < z_\omega$ and $z_n < z_{\omega+1}$ for all $n < \omega$. It then follows, by definition of c, that for all $n < \omega$ we have

$$z_n \in F_{z_\omega} \text{ and } z_n \in F_{z_{\omega+1}},$$

and therefore $F_{z_\omega} \cap F_{z_{\omega+1}}$ is infinite, contradicting condition (2) which we imposed on the construction of the sequence $\{F_\alpha\}_{\alpha < \omega_2}$. $\square$

We have shown that our colouring c has no homogeneous sets of the required sizes, and so our theorem is proved. $\square$

13.3 $\omega_1 \to (\omega_1, \alpha)^2$

Definition. A set A of ordinals is said to be *decomposable* if we can write $A = A_0 \cup A_1$ such that

$$otp(A) > otp(A_0), otp(A_1).$$

Otherwise, A is said to be *indecomposable*.

Note that in [23, section XIV.6], indecomposable ordinals are called *prime components* or *principal numbers of addition*.

Fact. *Indecomposable order types are powers of ω. In other words, A is indecomposable iff $otp(A) = \omega^\alpha$ for some ordinal α.*

Examples of indecomposable order types are

$$\omega_0 = 1, \omega^1 = \omega, \omega^2, \ldots, \omega^\omega, \omega^{\omega+1}, \ldots, \omega^{\omega_1} = \omega_1, \ldots.$$

Notation. For sets of ordinals A and B, we define

$$A < B \iff (\forall a \in A)(\forall b \in B)[a < b].$$

Fact. *If A is countably infinite and indecomposable, then we can write*

$$A = \bigcup_{n=0}^{\infty} A_n$$

where we have $A_0 < A_1 < \cdots < A_n < \cdots$, and each A_n is indecomposable.
[This can be proven by induction on the order type.]

Fact. *For every set of ordinals X, there is a finite decomposition*

$$X = X_0 \cup X_1 \cup \cdots \cup X_{n-1}$$

where we have $X_0 < X_1 < \cdots < X_{n-1}$ and each X_i is indecomposable.
[In the case X is indecomposable, simply set $n = 1$.]

Definition. An ultrafilter $\mathcal{U}$ on a set A is called *uniform* if $otp(B) = otp(A)$ for all $B \in \mathcal{U}$. [Is this standard? The textbooks define uniform based on cardinality, not order type.]

Fact. *If $otp(A) = \omega$, then an ultrafilter on A is uniform iff it is non-principal.*

Fact. *If A is indecomposable then there exists a uniform ultrafilter on A.*

The following variation of Solovay's Lemma will be useful:

Lemma. *Suppose we have a uniform ultrafilter $\mathcal{V}$ on $\mathbb{N}$, and suppose $\mathcal{Z} \subseteq \mathcal{V}$ has cardinality $< \mathfrak{m}$. Then there is an infinite set $W \subseteq \mathbb{N}$ such that $W \setminus Z$ is finite for all $Z \in \mathcal{Z}$.*

Proof. Let $\mathcal{A} = \{\mathbb{N}\}$, and define the set

$$\mathcal{B} = \{\mathbb{N} \setminus Z : Z \in \mathcal{Z}\}.$$

It is clear that we have $|\mathcal{B}| < \mathfrak{m}$. Also, for any finite subset $\mathcal{F} \subseteq \mathcal{B}$, it is clear that $\mathbb{N} \setminus \bigcup \mathcal{F}$ is a finite intersection of sets from $\mathcal{V}$, and so $\mathbb{N} \setminus \bigcup \mathcal{F}$ is in $\mathcal{V}$ (since $\mathcal{V}$ is closed under finite intersections) and is therefore infinite (since $\mathcal{V}$ is uniform).

We can now apply Solovay's Lemma. Solovay's Lemma gives us an infinite set $W \subseteq \mathbb{N}$ such that $W \cap B$ is finite for every $B \in \mathcal{B}$. But that means $W \setminus Z$ is finite for every $Z \in \mathcal{Z}$, and we have completed the proof. $\square$

Definition. Suppose we fix some arbitrary uniform ultrafilter $\mathcal{V}$ on $\mathbb{N}$. Suppose that for each $n \in \mathbb{N}$, A_n is a set of ordinals, and $\mathcal{U}_n$ is an ultrafilter on A_n. Suppose also that $A_n < A_{n+1}$ for each n, and that

$$A = \bigcup_{n=0}^{\infty} A_n.$$

We define the limit

$$\lim_{n \to \mathcal{V}} \mathcal{U}_n \subseteq \mathcal{P}(A)$$

by setting, for all $X \subseteq A$,

$$X \in \lim_{n \to \mathcal{V}} \mathcal{U}_n \iff \{n \in \mathbb{N} : X \cap A_n \in \mathcal{U}_n\} \in \mathcal{V}.$$

Fact. *The limit of uniform ultrafilters is a uniform ultrafilter.*

We will now define a particular uniform ultrafilter $\mathcal{U}_A$ to be associated with each countably infinite indecomposable set of ordinals A. We will carry out the construction recursively on the infinite indecomposable ordinals, as follows:

Start by fixing an arbitrary uniform ultrafilter $\mathcal{V}$ on $\mathbb{N}$, and let

$$\mathcal{U}_\omega = \mathcal{V}.$$

Now for any countably infinite indecomposable set of ordinals A with $otp(A) > \omega$, by a fact above we can write

$$A = \bigcup_{n=0}^{\infty} A_n$$

where we have $A_0 < A_1 < \cdots < A_n < \cdots$, and each A_n is indecomposable. By recursion we may assume that for each n we have already defined a particular uniform ultrafilter $\mathcal{U}_{A_n}$ on A_n. We then define

$$\mathcal{U}_A = \lim_{n \to \mathcal{V}} \mathcal{U}_{A_n}.$$

It is clear that $\mathcal{U}_A$ is a uniform ultrafilter on A.

Lemma. *Suppose A is a countably infinite indecomposable set of ordinals, and let $\mathcal{U}_A$ be defined as above. Then for every subset $\mathcal{X} \subseteq \mathcal{U}_A$ of cardinality $< \mathfrak{m}$ there is a $B \subseteq A$ such that $otp(B) = otp(A)$ and $B \setminus X$ is bounded in B for all $X \in \mathcal{X}$.*

Proof. The proof is by induction on $otp(A)$.

For the base case, suppose $otp(A) = \omega$. Then $\mathcal{U}_A$ is a uniform ultrafilter on $\mathbb{N}$, and so the previous lemma gives us our result.

We now assume $otp(A) > \omega$, and we assume that our lemma has been proven for order types smaller than that of A. By an earlier fact, we can write

$$A = \bigcup_{n=0}^{\infty} A_n$$

where we have $A_0 < A_1 < \cdots < A_n < \cdots$, and each A_n is indecomposable.

For each $X \in \mathcal{X}$, define the set

$$Z_X = \{n \in \mathbb{N} : X \cap A_n \in \mathcal{U}_{A_n}\}.$$

Since

$$X \in \mathcal{X} \subseteq \mathcal{U}_A = \lim_{n \to \mathcal{V}} \mathcal{U}_{A_n},$$

it follows by definition of $\lim_{n \to \mathcal{V}} \mathcal{U}_{A_n}$ that $Z_X \in \mathcal{V}$. Let

$$\mathcal{Z} = \{Z_X : X \in \mathcal{X}\} \subseteq \mathcal{V}.$$

Since $|\mathcal{Z}| < \mathfrak{m}$, we can apply our version of Solovay's Lemma to obtain an infinite set $W \subseteq \mathbb{N}$ such that $W \setminus Z_X$ is finite for all $X \in \mathcal{X}$.

Fix $n \in W$, and consider the family

$$\mathcal{X}_n = \{X \cap A_n : X \in \mathcal{X}\} \cap \mathcal{U}_{A_n}.$$

Clearly we have $\mathcal{X}_n \subseteq \mathcal{U}_{A_n}$ and $|\mathcal{X}_n| < \mathfrak{m}$. By the inductive hypothesis, there is a set $B_n \subseteq A_n$ such that $otp(B_n) = otp(A_n)$ and $B_n \setminus Y$ is bounded in B_n for all $Y \in \mathcal{X}_n$. This means $B_n \setminus X$ is bounded in B_n for all $X \in \mathcal{X}$ such that $X \cap A_n \in \mathcal{U}_{A_n}$.

Since B_n is countable, we can enumerate

$$B_n = \{b_{n,j}\}_{j < \omega}.$$

We then choose a subsequence $\{c_{n,k}\}_{k < \omega}$ as follows: For every $k < \omega$, let j_k be the smallest number j such that $b_{n,j} > c_{n,l}$ for all $l < k$, and then set $c_{n,k} = b_{n,j_k}$. In this way, we have chosen an increasing cofinal sequence

$$\{c_{n,k}\}_{k < \omega} \subseteq B_n.$$

For each $k < \omega$, we define the set

$$B_n^k = \{\alpha \in B_n : \alpha \le c_{n,k}\}.$$

It is clear that

$$B_n = \bigcup_{k=0}^{\infty} B_n^k,$$

and $B_n^k \subseteq B_n^{k+1}$ for each $k < \omega$. We also have that every set that is bounded in B_n is included in B_n^k for some k.

Now, fix $X \in \mathcal{X}$, and suppose $n \in W \cap Z_X$. We then have (by definition of Z_X) that $X \cap A_n \in \mathcal{U}_{A_n}$. It follows (by choice of B_n above) that $B_n \setminus X$ is bounded in B_n. This means we can choose some $k_{n,X} < \omega$ such that

$$B_n \setminus X \subseteq B_n^{k_{n,X}}.$$

Define a function $h_X : W \to \omega$ as follows: For $n \in W \cap Z_X$, let

$$h_X(n) = k_{n,X}.$$

For $n \in W \setminus Z_X$, define $h_X(n)$ arbitrarily.

Consider the collection

$$\mathcal{H} = \{h_X : X \in \mathcal{X}\}$$

of functions from W to ω. We know that $\mathfrak{m} \leq \mathfrak{b}$ (see [14]), where $\mathfrak{b}$ is the *bounding number*, that is the minimum cardinality of an unbounded family of functions. We have

$$|\mathcal{H}| = |\mathcal{X}| < \mathfrak{m} \leq \mathfrak{b},$$

so clearly the family $\mathcal{H}$ must be bounded. This means there is a function $h : W \to \omega$ such that h eventually dominates h_X for all X.

Define

$$B = \bigcup_{n \in W} \left(B_n \setminus B_n^{h(n)} \right),$$

and we will show that B satisfies the required conditions.

First, it is clear that

$$B = \bigcup_{n \in W} \left(B_n \setminus B_n^{h(n)} \right) \subseteq \bigcup_{n \in W} B_n \subseteq \bigcup_{n \in W} A_n \subseteq \bigcup_{n=0}^{\infty} A_n = A.$$

Next, notice that $otp(B) = otp(A)$.

Finally, fix some $X \in \mathcal{X}$. We know that h eventually dominates h_X, that is, there is some $m_X \in \mathbb{N}$ such that $h(n) > h_X(n)$ for all $n \in W$ with $n > m_X$. Since $W \setminus Z_X$ is finite, we can make sure to choose m_X large enough so that $W \setminus Z_X \subseteq m_X$. So for all $n \in W$ with $n > m_X$, we have $n \in W \cap Z_X$, and so

$$B_n \setminus X \subseteq B_n^{k_{n,X}} = B_n^{h_X(n)} \subseteq B_n^{h(n)},$$

and therefore

$$\left(B_n \setminus B_n^{h(n)} \right) \setminus X = \emptyset.$$

It follows that

$$B \setminus X = \bigcup_{n \in W} \left(B_n \setminus B_n^{h(n)} \right) \setminus X = \bigcup_{\substack{n \in W \\ n \leq m_X}} \left(B_n \setminus B_n^{h(n)} \right) \setminus X$$

and this set is clearly bounded in B, as any element of B_n with $n \in W$ and $n > m_X$ will be an upper bound for $B \setminus X$ in B.

So our set B satisfies the required conditions and the proof is complete. $\square$

Fact. *In the above lemma, we can in fact choose B such that $B \setminus X$ is finite, but we have not proven that.*

Fact. *The result of the above lemma remains true for any uniform ultrafilter on A, not just the $\mathcal{U}_A$ specifically constructed. But the proof of the general result would be more complicated, and we do not need the general result to prove our next theorem.*

Notation. For any given sets A and B, we define

$$A \otimes B = \big\{\{\alpha, \beta\} : \alpha \in A, \beta \in B, \alpha \neq \beta\big\} \subseteq [A \cup B]^2.$$

Theorem 87. *If $\mathrm{mm} > \omega_1$ then*

$$\omega_1 \to (\omega_1, \alpha)^2 \text{ for all } \alpha < \omega_1.$$

Proof. Fix a partition $[\omega_1]^2 = K_0 \cup K_1$. We need to show that for any countable ordinal α, there is either

1. an uncountable $X \subseteq \omega_1$ such that $[X]^2 \subseteq K_0$, or

2. a subset $W \subseteq \omega_1$ with order type α such that $[W]^2 \subseteq K_1$.

Since alternative (1) does not involve the ordinal α, we can rephrase the requirement as follows: We need to show that either

1. there is an uncountable $X \subseteq \omega_1$ such that $[X]^2 \subseteq K_0$, or

2. for any $\alpha < \omega_1$, there is a subset $W \subseteq \omega_1$ with order type α such that $[W]^2 \subseteq K_1$.

Suppose alternative (1) is false. We will prove that alternative (2) must hold, thereby proving the theorem.

We need a stronger form of (2) to use as our inductive hypothesis. In fact we will prove the following statement, by induction on $\alpha < \omega_1$:

$(IH)_\alpha$: For every uncountable $X \subseteq \omega_1$, there is a subset $W \subseteq X$ with order type α and an uncountable subset $Y \subseteq X$ such that $W < Y$ and $[W]^2 \subseteq K_1$ and $W \otimes Y \subseteq K_1$.

[Note that in $(IH)_\alpha$, we do *not* claim that $[Y]^2 \subseteq K_1$.]

Fix a countable ordinal α, and assume that $(IH)_\gamma$ is true for all $\gamma < \alpha$. We must prove $(IH)_\alpha$. Fix an uncountable set $X \subseteq \omega_1$.

If α is decomposable, then we have $\alpha = A_0 \cup A_1$ with $otp(A_0), otp(A_1) < \alpha$. We now apply $(IH)_{otp(A_0)}$ on the set X, and we obtain a subset $W_0 \subseteq X$

with $otp(W_0) = otp(A_0)$ and an uncountable subset $Y_0 \subseteq X$ such that $W_0 < Y_0$ and $[W_0]^2 \subseteq K_1$ and $W_0 \otimes Y_0 \subseteq K_1$. Since Y_0 is uncountable, we then apply $(IH)_{otp(A_1)}$ on the set Y_0, and we obtain a subset $W_1 \subseteq Y_0$ with $otp(W_1) = otp(A_1)$, and an uncountable subset $Y_1 \subseteq Y_0$, such that $W_1 < Y_1$ and $[W_1]^2 \subseteq K_1$ and $W_1 \otimes Y_1 \subseteq K_1$. Let $W = W_0 \cup W_1$ and let $Y = Y_1$. We have

$$otp(W) = otp(W_0 \cup W_1) = otp(A_0) + otp(A_1) = \alpha.$$

We know Y is uncountable and clearly $W < Y$ (since $W_0 < Y_0$ and $W_1 < Y_1$). We have

$$\begin{aligned}
[W]^2 &= [W_0 \cup W_1]^2 \\
&= [W_0]^2 \cup (W_0 \otimes W_1) \cup [W_1]^2 \\
&\subseteq [W_0]^2 \cup (W_0 \otimes Y_0) \cup [W_1]^2 \\
&\subseteq K_1.
\end{aligned}$$

Also,

$$\begin{aligned}
W \otimes Y &= (W_0 \cup W_1) \otimes Y \\
&= (W_0 \otimes Y) \cup (W_1 \otimes Y) \\
&\subseteq (W_0 \otimes Y_0) \cup (W_1 \otimes Y_1) \\
&\subseteq K_1.
\end{aligned}$$

So we have proved $(IH)_\alpha$ in the case where α is decomposable.

We may now assume that α is indecomposable. Define a set mapping

$$F : X \to [X]^{\leq \aleph_0}$$

by setting, for all $\eta \in X$,

$$F(\eta) = \{\xi \in X : \xi < \eta \text{ and } \{\xi, \eta\} \in K_1\}.$$

Since we are assuming $\mathfrak{mm} > \omega_1$, we know from Theorem 56 that the Sparse Set Mapping Property (SMP) holds. Applying SMP to F, we see that either F is not sparse, or it admits an uncountable F-free set $W \subseteq X$.

Suppose there is an uncountable F-free set $W \subseteq X$. This means $F(\eta) \cap W \subseteq \{\eta\}$ for every $\eta \in W$. But it is clear from the definition of F that $\eta \notin F(\eta)$, so we have $F(\eta) \cap W = \emptyset$ for every $\eta \in W$. This means that for every $\eta \in W$, there is no $\xi < \eta$ in W such that $\{\xi, \eta\} \in K_1$. But this means $[W]^2 \subseteq K_0$, and since W is uncountable this contradicts our assumption that (1) is false.

The remaining possibility is that F is not sparse. This means that for every proper σ-ideal $\mathcal{I}$ on X, there is $Z \in \mathcal{I}^+(= \mathcal{P}(X) \setminus \mathcal{I})$ such that for every $Z_0 \in [Z]^{\leq \aleph_0}$ and every $Y \in \mathcal{I}$ there is a finite $B \subseteq X \setminus Y$ such that $Z_0 \subseteq \bigcup_{x \in B} F(x)$.

Notice that $[X]^{\leq \aleph_0}$ is a proper σ-ideal on X. So we can fix $Z \in \mathcal{P}(X) \setminus [X]^{\leq \aleph_0}$ (that is, an uncountable $Z \subseteq X$) such that for every countable $Z_0 \subseteq Z$ and every countable $Y \subseteq X$ there is a finite $B \subseteq X \setminus Y$ such that $Z_0 \subseteq \bigcup_{x \in B} F(x)$.

We will recursively construct a sequence $\{a_\xi\}_{\xi < \omega_1}$ of finite subsets of X, as follows: Fix $\xi < \omega_1$, and suppose we have defined a_η for $\eta < \xi$. Let $Y = \bigcup_{\eta < \xi} a_\eta$. Clearly $Y \subseteq X$ is countable, as it is a countable union of finite sets. Since ξ is a countable ordinal, $Z \cap \xi$ is a countable subset of Z, so by choice of Z in the previous paragraph, we can find a finite $B \subseteq X \setminus Y$ such that

$$Z \cap \xi \subseteq \bigcup_{x \in B} F(x).$$

Let a_ξ be this B, completing the recursive construction.

So the collection $\{a_\xi\}_{\xi < \omega_1}$ consists of pairwise disjoint finite subsets of X such that for all $\eta \leq \xi < \omega_1$ we have

$$Z \cap \eta \subseteq Z \cap \xi \subseteq \bigcup_{x \in a_\xi} F(x).$$

Since α is countable and indecomposable, we can write

$$\alpha = \sum_{n=0}^{\infty} \alpha_n$$

where each α_n is an indecomposable ordinal such that $\alpha_n < \alpha$.

We now recursively construct sequences $\{W_n\}_{n < \omega}$ and $\{Y_n\}_{n < \omega}$ of subsets of Z, as follows: For $n = 0$, we apply $(IH)_{\alpha_0}$ on the uncountable set Z, and we obtain a subset $W_0 \subseteq Z$ with order type α_0 and an uncountable subset $Y_0 \subseteq Z$ such that $W_0 < Y_0$ and $[W_0]^2 \subseteq K_1$ and $W_0 \otimes Y_0 \subseteq K_1$.

For $n > 0$, given uncountable $Y_{n-1} \subseteq Z$ already constructed, we apply $(IH)_{\alpha_n}$ on the set Y_{n-1}, and we obtain a subset $W_n \subseteq Y_{n-1}$ of order type α_n and an uncountable subset $Y_n \subseteq Y_{n-1}$ such that $W_n < Y_n$ and $[W_n]^2 \subseteq K_1$ and $W_n \otimes Y_n \subseteq K_1$.

We let

$$W = \bigcup_{n=0}^{\infty} W_n.$$

Clearly $otp(W) = \alpha$ and $[W]^2 \subseteq K_1$. If we needed to prove only (2) for α, we could stop here; but we need to prove the stronger $(IH)_\alpha$.

Since $W \subseteq Z$ is countable we have $W \subseteq Z \cap \eta_0$ for some countable ordinal η_0. So for each countable ordinal $\xi \geq \eta_0$ we have

$$W \subseteq Z \cap \eta_0 \subseteq \bigcup_{x \in a_\xi} F(x),$$

and therefore

$$W = \bigcup_{x \in a_\xi} F(x) \cap W.$$

We have thus written W as a union of finitely many subsets, and so it follows that one of the subsets must be a member of the ultrafilter $\mathcal{U}_W$. In other words, we can choose some $\beta_\xi \in a_\xi$ such that $F(\beta_\xi) \cap W \in \mathcal{U}_W$.

The collection

$$\{F(\beta_\xi) \cap W\}_{\eta_0 \le \xi < \omega_1} \subseteq \mathcal{U}_W$$

has size $\omega_1 < \mathfrak{mm} \le \mathfrak{m}$, and so by the lemma there is a $B \subseteq W$ such that $otp(B) = otp(W) = \alpha$ and $B \setminus (F(\beta_\xi) \cap W)$ is bounded in B for each countable $\xi \ge \eta_0$.

For each countable $\xi \ge \eta_0$, we have $B \setminus F(\beta_\xi)$ is bounded in B, so we can choose some $\delta_\xi \in B$ that bounds $B \setminus F(\beta_\xi)$. Since B is countable and there are uncountably many such ordinals ξ, by the Pigeonhole Principle there must be some $\delta \in B$ and some uncountable set $\Gamma \subseteq \omega_1 \setminus \eta_0$ such that $\delta = \delta_\xi$ for all $\xi \in \Gamma$. Define

$$\tilde{B} = B \setminus \delta,$$

and we then have $\tilde{B} \subseteq F(\beta_\xi)$ for all $\xi \in \Gamma$.

Define

$$Y = \{\beta_\xi : \xi \in \Gamma\}.$$

We then have $\tilde{B} \subseteq F(y)$ for all $y \in Y$.

We will now show that $\tilde{B}$ and Y are the sets required to satisfy $(IH)_\alpha$:

Since $otp(B) = \alpha$ is indecomposable, the 'tail' $\tilde{B}$ of B must have the same order type as B, that is, $otp(\tilde{B}) = \alpha$, as required.

For any $\xi_1 \ne \xi_2 \in \Gamma$, we know by construction that $a_{\xi_1} \cap a_{\xi_2} = \emptyset$, and so $\beta_{\xi_1} \ne \beta_{\xi_2}$. It follows that Y is uncountable since Γ is uncountable. Clearly $Y \subseteq X$ since $a_\xi \subseteq X$ for each ξ.

We also have

$$[\tilde{B}]^2 \subseteq [B]^2 \subseteq [W]^2 \subseteq K_1.$$

For any $b \in \tilde{B}$ and any $y \in Y$, we have

$$b \in \tilde{B} \subseteq F(y),$$

and so by definition of F, it follows that $b < y$ and $\{b, y\} \in K_1$. It follows that $\tilde{B} < Y$ and $\tilde{B} \otimes Y \subseteq K_1$.

We have verified all the conditions and so we have proven $(IH)_\alpha$. This completes the induction and so our theorem is proven. $\qquad\square$

Chapter 14

Five Cofinal Types

14.1 Tukey reductions and cofinal equivalence

In order to understand the structure of a topological space, sometimes it is necessry to study it in terms of the clousure operator and its relationship with the "convergent sequences". When we deal with spaces like $X = [0,1]^{\omega_1}$, we may have "transfinite sequences" of the form $f : \omega_1 \to X$, so we introduce the notion of a generalized sequence and convergence (Moore-Smith convergence.)

Definition. A partially ordered set $(D, \leq_D)$ is directed iff for every x, y in D there is $z \in D$ such that $x \leq_D z$ and $y \leq_D z$.

Definition. Let X be a topological space and D a directed set, we say that a map $f : D \to X$ is a net. A net $f : D \to X$ converges to a point $x \in X$ iff for every U open neighborhood containing x there exists $d_0 \in D$ such that $f(d) \in U$ whenever $d \geq d_0$.

So, it is essential to know the different kind of points we have in a given topological space, i.e., the directed sets we need in order to define the clousure operator. We will focuse in the those directed sets of size at most $\aleph_1$. The following notions were introduced by Tukey [34], and they allow us to know when can we use indistinctly two different directed sets to define a net.

Definition. Let E and D be directed sets. We say that $D \geq E$ iff there is a map $f : D \to E$ such that for every $e \in E$ there is $d \in D$ such that $f(d') \geq e$ whenever $d' \geq d$ (we say in this case that f is a convergent map).

Definition. Let D and E be directed sets. We say that $g : E \to D$ is Tukey if g maps unbounded subsets of E into unbounded subsets of D.

Lemma. *If D and E are directed sets, then $D \geq E$ iff there is a Tukey map $g : E \to D$.*

Proof. Suppose first $D \geq E$, i.e., there is $f : D \to E$ a convergent map. Define $g : E \to D$ as follows: let $e \in E$ and choose $g(e) \in D$ such that $f(d') \geq e$ whenever $d' \geq g(e)$, this $g(e)$ exists because f is a convergent map. Notice that the set $\{x \in E : g(x) \leq d\}$ is bounded in E by $f(d)$, therefore g maps unbounded subsets of E into unbounded subsets of D. Conversely, suppose $g : E \to D$ is a Tukey map. Define $f : D \to E$ such that for every $d \in D$, $f(d)$ is an upper bound for $X = \{x \in E : g(x) \leq d\}$. To prove that f is convergent, let $e \in E$, for each $d' \geq g(e)$ consider the set $\{x \in E : g(x) \leq d'\}$, since $e \in X$ and $f(d')$ is an upper bound for X, we have that $f(d') \geq e$, hence f is a convergent map. $\qquad\square$

Definition. Let D and E be directed sets. We say that D and E are cofinally equivalent($D \equiv E$) iff $D \geq E$ and $E \geq D$.

Theorem 88 (Tukey). *$D \equiv E$ iff there is a directed set $(X, \leq_X)$ such that both D and E are isomorphic to a cofinal subset of X.*

Proof. Assume that $D \equiv E$. then we can find convergent maps $f : D \to E$ and $g : E \to D$ such that if $d \in D$ and $e \in E$, $d \geq_D g(e)$implies that $f(d) \geq_E e$ and $e \geq_E f(d)$ implies $g(e) \geq_D d$. Let $X = D \cup E$. (assuming that $E \cap D = \emptyset$). Define $\leq_X$ as follows: $\leq_X$ restricted to D equal to $\leq_D$ and $\leq_X$ restricted to E equal to $\leq_E$. Let $d \in D$ and $e \in E$. Set $e \leq_X d$ iff $g(e') \leq_D d$ for some $e' \geq_E e$ and $d \leq_X e$ iff $f(d') \leq_E e$ for some $d' \geq_D d$, then $\leq_X$ is a quas-iordering on X, by taking the eqivalence clases $\{[x] : x \in X\}$ with respect to the equivalence relation induced by $\leq_X$, and denote this set by $\bar{X}$, we have the E and D are isomorphic to cofinal subsets of $\bar{X}$ by identifying e with $[e]$ and d with $[d]$, for each $e \in E$ and $d \in D$. $\quad\square$

Then we have that the relation "$\equiv$" among directed sets is an equivalence relation. The equivalence classes under this relation are called **cofinal types**.

14.2   Directed posets of cardinality at most $\aleph_1$

Observe that the only countable cofinal types are either 1 and ω. Some examples of cofinal types of size $\leq \omega_1$ and cofinality ω_1 are $\omega_1, \omega \times \omega_1$, $[\omega_1]^{<\omega}$. We will prove in fact (assuming $\mathfrak{mm} > \omega_1$) that these are the only cofinal types of cardinality at most $\aleph_1$.

Theorem 89. *If* $\mathfrak{mm} > \omega_1$, *then* $1, \omega, \omega_1, \omega \times \omega_1$ *and* $[\omega_1]^{<\omega}$ *are the only cofinal types of directed sets of cardinality at most* $\aleph_1$.

Proof. Let $(D, \leq_D)$ be a directed set of cardinality at most $\aleph_1$. We have then two cases, either D has cofinality $< \omega_1$ or D has cofinality ω_1. In the first case if D has maximal element, we have that D is cofinally equivalent to 1 otherwise D contains a cofinal chain of type ω in which case D is cofinally equivalent to ω. In the case D has cofinality equal to ω_1 we proceed as follows. Observe that we may assume that $D = (\omega_1, \leq_D)$, where $\alpha < \beta$, whenever $\alpha \leq_D \beta$. (Any ω_1-cofinal subsequence of D is cofinaly equivalent to D.)

Consider the family

$$\mathcal{I} = \{X \subseteq D : (\forall X_0 \in [X]^\omega)(\exists d \in D)(\forall s \in X_0)(x \leq_D d)\}.$$

Using the fact that D is directed, we can easily prove that $\mathcal{I}$ is an ideal. We have now two cases to be considered.

Case 1 D can not be written as a countable union of elements of the ideal $\mathcal{I}$, (this condition is equivalent to say that $\omega \times \omega_1 \not\geq D$).

In this case we have that $\mathcal{I}$ is a proper σ ideal. We will work to prove that in this case $D \equiv [\omega_1]^{<\omega}$.

Consider the partial order consisting of all the pairs $p = (X_p, \mathcal{N}_p)$, where

(a) $X_p \in [D]^{<\omega}$

(b) $\mathcal{N}_p$ is a finite ϵ-chain of countable elementary submodels of (H_{ω_2}, ϵ)

(c) $(\forall x \neq y \in X_p)(\exists N \in \mathcal{N}_p)(|N \cap \{x, y\}| = 1)$

(d) $(\forall N \in \mathcal{N}_p) (\forall x \in X_p \smallsetminus N)(\forall A \in \mathcal{I} \cap N)(x \notin A)$
we set $p \leq q$ iff

(e) $X_p \supseteq X_q$, $\mathcal{N}_p \supseteq \mathcal{N}_q$ and

(f) $(\forall x \in X_q) (\forall y \in X_p \smallsetminus X_q)(y \not\leq_D x)$.

Consider the map $F : D \to [D]^{\leq \omega}$ defined by $F(d) = \{x \in D : x \leq d\}$. Observe that F is $\mathcal{I}$-sparse. Acording to the definition of sparsesness, given $Z \in \mathcal{I}^+$ we need to find $Z_0 \in [Z]^{\leq \omega}$ and $H \in \mathcal{I}$ such that $X_0 \not\subseteq \bigcup_{d \in A} F(d)$ for all $A \in D \smallsetminus H$. This is obviously true, since given $Z \notin \mathcal{I}$, there is Z_0 a countable subset of Z_0 without an upper bound in D, and if $Z_0 \subseteq \bigcup_{d \in A} F(a)$ for $A \in D \setminus H$ with $H \in \mathcal{I}$, by definition of F it would imply that Z_0 is bounded in D.

Using the proof that $\mathfrak{mm} > \omega_1$ implies **SMP**, we proof that $\mathcal{P}$ is proper. Now for each $z \in D$ we define

$$\mathcal{D}_z = \{p \in \mathcal{P} : (\exists x \in X_p)(z \leq x)\}$$

Claim. *For each $z \in D$ the set $\mathcal{D}_z$ is open dense.*

Proof. Let $z \in D$. By assumption the set $\{y : y \geq z\} \notin \mathcal{I}$. Then given $q \in \mathcal{P}$, we find N a countable elementary submodel of (H_{ω_2}, ϵ) such that $z, q, D \in N$ and find $x \in D_z \setminus N$ such that $x \notin A$ for every $A \in \mathcal{I} \cap N$. Then by definition of $\mathcal{P}$, we have that $p = (X_p \cup \{x\}, \mathcal{N}_p \cup \{N\})$ is an element of $\mathcal{D}_z$ extending p. $\square$

Using $>\omega_1$, we find $\mathcal{G} \subseteq \mathcal{P}$ a generic filter intersecting $\mathcal{D}_z$ for every $z \in D$. Define

$$X = \bigcup_{p \in \mathcal{G}} X_p$$

Then x is a cofinal subset of D with the property that for each $x \in X$ the set $\{y \in X : y \leq x\}$ is finite. This implies that every infinite subset of X is unbounded in D. Hence if we consider $f : [\omega_1]^{<\omega} \to X$ any one to one function, we have that it is in particular unbounded. Therefore $[\omega_1]^{<\omega} \leq D$ and in consequence $D \equiv [\omega_1]^{<\omega}$.

Case 2 D can be written as $D = \bigcup_{n=0}^{\infty} D_n$, where $D_n \in \mathcal{I}$ for each $n \in \mathbb{N}$.

In this case we have that $D \leq \omega_1 \times \omega$. We have two cases

Case 2.1 : D can be covered by finitely many subsets of the ideal $\mathcal{I}$. This is equivalent to say that every countable subset of D has an upper bound on D, so we can get a cofinal sequence of order type ω_1, in consequence D is equivalent to ω_1.

Case 2.2 : D cannot be covered by finetely many elements from $\mathcal{I}$. We know D can be covered by countably many sets from $\mathcal{I}$ and therefore $D \leq \omega \times \omega_1$. We will define $f : D \to \omega \times \omega_1$ a Tukey map (so we will have that $D \geq \omega \times \omega_1$ and in consequence $D \equiv \omega \times \omega_1$). Let $D = \bigcup_{n \in \omega} D_n$, where D_n are disjoint element from $\mathcal{I}$. We may assume that each D_n is uncountable. (If we remove the countable parts the remaining set is still equivalent to D.) Observe that we can also take $D'_n s$ downwards closed and such that $D_n \subseteq D_{n+1}$ for each $n \in \mathbb{N}$ nad $D_{n+1} \setminus D_n$ is uncountable. For each $n \in \mathbb{N}$, choose $h_n : \omega_1 \to D_{n+1} \setminus D_n$ a one to one function. Define $f : \omega \times \omega_1 \to D$ by $f(n, \alpha) = h_n(\alpha)$. We claim that f is Tukey. Let $X \subseteq \omega \times \omega_1$ unbounded. If the first proyection of X is infinite, then $f[X]$ is unbounded, since in this case a possible bound for $f[X]$ have to be a member of D_n for some $n \in \mathbb{N}$. But a member of D_n cannot never dominate a member of $D_{m+1} \setminus D_m$ for $m \geq n$, (each D_n is downwards closed). Now if the first proyection of X is finite, we have that every countable subset of X has an upper bound on $\omega \times \omega_1$, so there is $n \in \mathbb{N}$ such that $X_n = (\{n\} \times \omega_1) \cap X$ is uncountable and $f(x) = h_n(x)$. So $f[X_n]$ is an uncountable subset of $D_{n+1} \setminus D_n$ and therefore unbounded on D. So we conclude that $D \equiv \omega \times \omega_1$.

$\square$

A similar result holds for all posets of cardinality at most $\aleph_1$ though the list of possible Tukey types is countably infinite and the arguments to achieve this are considerably deeper (see [29]). To state this result, let

$$D_0 = 1, D_1 = \omega, D_2 = \omega_1, D_3 = \omega \times \omega_1 \text{ and } D_4 = [\omega_1]^{<\omega}.$$

Let $\bigoplus_{i \in I} P_i$ be our natation for the natural disjoint incomparable sum of posets P_i $(i \in I)$. If $P_i = P$ for all $i \in I$ and of I has cardinality κ, then $\bigoplus_{i \in I} P_i$ is denoted by $\kappa \cdot P$ or κP.

Theorem. Assuming $\mathfrak{mm} > \omega_1$, every poset of size $\aleph_1$ is Tukey-equivalent to one from the list:

(a) $\bigoplus_{i<5} n_i D_i$ $(i < 5, n_i < \omega)$,

(b) $\aleph_0 \cdot 1 \oplus \bigoplus_{i=2}^{4} n_i D_i$ $(2 \leq i < 5, n_i < \omega)$,

(c) $\aleph_0 \cdot \omega_1 \oplus n_4 [\omega_1]^{<\omega}$ $(n_4 < \omega)$,

(d) $\aleph_0 \cdot [\omega_1]^{<\omega}$,

(e) $\aleph_1 \cdot 1$.

14.3   Directed sets of cardinality continuum

The result of the previous section does not generalize to directed sets of size continuum. The easiest ways to see this is to consider directed sets of the form

$$\mathcal{K}(X) = \{K \subseteq X : K \text{ is compact}\},$$

for X running in sum class of topological spaces. In fact, the easiest is simply to consider the family

$$\{\mathcal{K}(X) : X \text{ is a subspace of } \omega_1\}$$

of directed sets of size at most continuum. This follows from the following fact.

Theorem 90. *For uncountable $X, Y \subseteq \omega_1$, the inequality $\mathcal{K}(X) \leq \mathcal{K}(Y)$ implies that $Y \setminus X$ is a nonstationary subset of ω_1.*

Proof. Suppose that $Y \setminus X$ is stationary and consider a map $f : \mathcal{K}(X) \to \mathcal{K}(Y)$. Choose a countable elementary submodel M of H_{ω_2} containing X, Y and f such that $\delta = M \cap \omega_1$ belongs to $Y \setminus X$. Using the fact that X is uncountable and the elementarity of M we can build a strictly increasing sequence ξ_n $(n < \omega)$ of elements of $X \cap \delta$ such that

1. $\sup_{n < \omega} \xi_n = \delta$.

2. for all $\beta < \delta$, either $\beta \in f(\{\xi_n\})$ for some $n < \omega$, or there is $\alpha < \beta$ such that $(\alpha, \beta] \cap f(\{\xi_n\}) = \emptyset$ for all $n < \omega$.

    Let

$$K = \{\delta\} \cup \overline{\bigcup_{n < \omega} f(\{\xi_n\})}.$$

Then K is a compact subset of Y which bounds the seguence $f(\{\xi_n\})$ $(n < \omega)$. On the other hand, the sequence $\{\xi_n\}$ $(n < \omega)$ is not bounded in $\mathcal{K}(X)$ since it accumulates to $\delta \notin X$. So, f is not a Tukey map. $\qquad\square$

Corollary 91. *Suppose that $\mathcal{F}$ is a family[1] of subsets of ω_1 such that $X \setminus Y \in \text{stat}(\omega_1)$ whenever $X \neq Y$ are chosen from $\mathcal{F}$. Then the corresponding family $\mathcal{K}(X)$ $(X \in \mathcal{F})$ is a family of directed sets representing different cofinal types.*

Similar construction could be done using the families $\mathcal{K}(X)$ of compact subsets of separable metric spaces X. However it is much more interesting to investigate the cofinal types and Tukey reductions between definable

---

[1]Clearly, there is such a family of cardinality $2^{\aleph_1}$.

directed sets. There are some important examples of such directed sets such as the following:

$$\mathbb{N}^{\mathbb{N}}, \ell_1, \mathrm{NWD}, etc.$$

Here is a typical result from this theory.

Theorem 92. *For a metrizable space X the inequality $\mathcal{K}(X) \leq \mathbb{N}^{\mathbb{N}}$ holds if and only if X is Polish.*

Chapter 15

Five Linear Orderings

15.1 Basis problem for uncountable linear orderings

This problem asks for a finite list $\mathcal{K}$ of uncountable linear orderings with the property that an arbitrary linear ordering L is countable if and only if $K \not\leq L$ for all $K \in \mathcal{K}$. Note that if such a basis $\mathcal{K}$ exists and if it is moreover a minimal such basis, then every ordering K in $\mathcal{K}$ will have to be rather canonical or at least *minimal* as an uncountable linear ordering, i.e., $K \leq L$ for every uncountable linear subordering L of K. Clearly, ω_1 and ω_1^* will be members of any basis for the uncountable linear orderings. Another member of the basis is found by considering the following notion.

Definition. A linearly ordered set L is *separable* if it contains a countable dense subset D with the property that for all $x <_L y$ in L there is $d \in D$ such that $x \leq_L d \leq_L y$.

Clearly, for every uncountable separable linear ordering L there is an uncountable set of reals X such that $X \leq L$. It turns out that under $\mathrm{mm} > \omega_1$ every pair of separable linear orderings of cardinality $\aleph_1$ are equivalent. So in particular, the class of uncountable separable linear orderings has a one-element basis. We shall prove this in the firs section of this Chapter.

15.2 Separable linear orderings

The purpose of this section is to prove the following result.

Theorem 93. *Assuming* $\mathrm{mm} > \omega_1$, *every set of reals of size* $\aleph_1$ *embeds into any uncountable separable linear ordering.*

Proof. Fix a set of reals B of cardinality $\aleph_1$ and let X be an arbitrary uncountable set of reals. We shall define a stationary preserving poset $\mathcal{P}$ which forces a strictly increasing map $f : B \to X$. The $\mathcal{P}$ will have its side conditions not finite $\in$-chains of countable elementary submodels of H_{c^+} but rather than finite *continuous* $\in$-chains of such submodels, i.e, mappings of the form $N : D \to [H_{c^+}]^\omega$ such that

1. D is a finite subset of ω_1.

2. $N(\xi) \prec H_{c^+}$ for all $\xi \in D$.

3. $N(\xi) \in N(\eta)$ for $\xi < \eta$ in D.

4. There is a continuous $\in$-chain $(N_\xi : \xi < \omega_1)$ of countable elementary submodels of H_{c^+} such that $N_\xi = N(\xi)$ for all $\xi \in D$.

Thus, let $\mathcal{P}$ be the collection of all pairs $p = (f_p, N_p)$ such that

(1) f_p is a finite strictly increasing partial function from B into X.

(2) N_p is a finite continuous $\in$-chain of countable elementary submodels of H_{c^+} that contain B and X.

(3) The range of f_p is separated by N_p, i.e., for every pair of different elements of $\text{rang}(f)$ there is a model of N_p containing one but not the other element.

(4) For every $x \in \text{dom}(f)$ there is $\xi \in \text{dom}(N_p)$ such that $x \in N_p(\xi)$ but $f_p(x) \notin N_p(\xi)$ but there is no such a ξ in $\text{dom}(N_p)$ that is a limit ordinal.

We order $\mathcal{P}$ by coordinatewise inclusions.

Claim 93.1. $\mathcal{P}$ *is stationary set preserving.*

Proof. Let $(C_p : p \in \mathcal{P})$ be a given $\mathcal{P}$-club. Let E be a stationary subset of ω_1, and $\bar{p} \in \mathcal{P}$. We need to find p extending $\bar{p}$ such that $C(p) \cap E \neq \emptyset$. Choose a countable elementary submodel M of $H_{(2^c)^+}$ containing all these objects such that $\delta = M \cap \omega_1 \in E$. Let

$$p = (f_{\bar{p}}, N_{\bar{p}} \cup \{(\delta, M \cap H_{c^+})\}).$$

Note that p belongs to $\mathcal{P}$. Choose q extending p such that $C(q) \setminus \delta \neq \emptyset$. It suffices to show that δ belongs to $C(q)$. Otherwise, we can find $r \leq q$ and $\gamma < \delta$ such that

$$(\forall s \leq r) \ C(s) \cap (\gamma, \delta] = \emptyset.$$

Increasing γ if necessary, we assume that $\gamma > \max(\text{dom}(N_r) \cap \delta)$. Let $\mathcal{X}$ be the collection of all conditions x of $\mathcal{P}$ that end-extend $r \cap M$ and realize the same type as r over the parameters γ and $r \cap M$. We shall prove by induction on the cardinality of $\mathrm{dom}(f_r) \setminus M$ that there is $\bar{r} \in \mathcal{X} \cap M$ compatible with r which will be the desired contradiction, since for any common extension s of r and $\bar{r}$, we have that $C(s) \cap (\gamma, \delta] \neq \emptyset$, contradicting the choice of r. For the inductive step from k to $k+1$ we assume that $|\mathrm{dom}(f_r) \setminus M| = k+1$ and assume the inductive hypothesis at k. Note that such $\bar{r}$ will be compatible with r as long as $f_{\bar{r}} \cup f_r$ is an increasing function. Let $a \in \mathrm{dom}(f)_r$ be such that $f_r(a)$ has the largest model, call it $N(a)$, in N_r to which it does not belong. Note that

$$(\mathrm{dom}(f_r) \cup \mathrm{range}(f_r)) \setminus \{f_r(a)\} \subseteq N(a).$$

We enumerate $\mathrm{dom}(f_r) \cup \mathrm{range}(f_r) \setminus M$ as $(a_0^r, ..., a_{2k+1}^r)$ such that $(a_0^r, ..., a_k^r)$ enumerates $\mathrm{dom}(f_r) \setminus M$ with $a_0^r = a$ and $(a_{k+1}^r, ..., a_{2k+1}^r)$ enumerates $\mathrm{range}(f_r) \setminus M$ according the $\in$-ordering of the models of $N_r \setminus M$ that separate them. Note that $a_{2k+1}^r = f_r(a)$. Since the conditions in $\mathcal{X}$ are isomorphic, every $x \in \mathcal{X}$ has its own $2k+2$-tuple $(a_0^x, ..., a_k^x, a_{2k+1}^x)$ that enumerates

$$\mathrm{dom}(f_x) \cup \mathrm{range}(f_x) \setminus (\mathrm{dom}(f_r) \cup \mathrm{range} f_r) \cap M$$

in the similar way. Let H be the corresponding subset of $\mathbb{R}^{2k}$ and let G be its projection to the first $2k-1$ coordinates. Let $H_0 \in M$ be a countable dense subset of H and let G_0 be its projection to the first $2k+1$ coordinates. Since H_0 is dense and since we have the induction hypothesis on $\mathrm{dom}(f_r) \setminus M$. For every open neighborhood U of $(a_0^r, ..., a_{2k+1}^r)$ there is $(a_0^x, ..., a_{2k+1}^x) \in U \cap H_0$ such that

$$(f_r \setminus \{(a_0^r, a_{2k+1}^r)\}) \cup (f_x \setminus \{(a_0^x, a_{2k+1}^x)\})$$

is an increasing function. We use the notation $(a_0^r, ..., a_{2k}^r)\rho(a_0^x, ..., a_{2k}^x)$ as a shorthand for the relation between these two $(2k+1)$-tuples. Running over all such neighborhoods U of $(a_0^r, ..., a_{2k+1}^r)$ the corresponding a_0^x's will approach a_0^r either from the left, or right, or both. We may assume that the first alternative happens. Then for every such $(a_0^r, ..., a_{2k+1}^r) \in U \cap H_0$ with $a_0^x < a_0^r$ we have that either $a_{2k+1}^x < a_{2k+1}^r$ or $a_{2k+1}^x > a_{2k+1}^r$. If the first alternative happens, we have reached our goal of finding $x \in \mathcal{X} \cap M$ compatible with r reaching thus the desired contradiction. So we are left to analyzing the case that $a_{2k+1}^x > a_{2k+1}^r$ for every $(a_0^x, ..., a_{2k+1}^x) \in U \cap H_0$ with $a_0^x < a_0^r$. It follows that

$$a_{2k+1}^r = \inf\{a_{2k+1}^x : (a_0^r, ..., a_{2k+1}^r) \in H_0, a_0^x < a_0^r, (a_0^r, ..., a_{2k}^r)\rho(a_0^x, ..., a_{2k}^x)\}.$$

Note that this equation defines a partial function g from a subset of $\mathbb{R}^{2k-1}$ into $\mathbb{R}$ such that $g(a_0^r, ..., a_{2k}^r) = a_{2k+1}^r$. However, this leads to a contradiction since, clearly, g belongs to the model $N(a)$ which separates $a_{2k+1}^r = f_r(a)$ from the rest of the points in the domain and range of f_r. This finishes the proof. $\qquad\square$

For $b \in B$, let
$$\mathcal{D}_b = \{p \in \mathcal{P} : b \in \mathrm{dom}(f_p)\}.$$
It is easily seen that each $\mathcal{D}_b$ is a dense-open subset of $\mathcal{P}$ so an filter meeting all these sets will give us the desired strictly increasing map from B into X. This finishes the proof.

□

Remark. The side conditions in the above proof are introduced as they will show up in some other applications below. What we want to point out here that for this particular application the following variant is perhaps more to the point. Thus, it is more up to the point to put the poset for Theorem 93 to be the set of triples $p = (f_p, \mathcal{M}_p, \mathcal{N}_p)$ such that

(1) f_p is a finite strictly increasing partial function from B into X.

(2) $\mathcal{N}_p$ is a finite $\in$-chain of countable elementary submodels of $H_{\mathfrak{c}^+}$ that contain B and X.

(3) $\mathcal{M}_p \subseteq \mathcal{N}_p$ and for every $M \in \mathcal{M}_p$ there is a countable elementary submodel $\bar{M}$ of $H_{(2^{\mathfrak{c}})^+}$ such that $M = \bar{M} \cap H_{\mathfrak{c}^+}$,

(4) The range of f_p is separated by $\mathcal{N}_p$,

(5) For every $x \in \mathrm{dom}(f)$ there is $N \in \mathcal{N}_p$ such that $x \in N$ but $f_p(x) \notin N$ but there is no $M \in \mathcal{M}_p$ that separates x and $f_p(x)$.

It should be clear that the above argument works for this version as well with esentially no extra changes. For example, at the beginning of the proof of Claim 93.1, the condition
$$p = (f_{\bar{p}}, \mathcal{M}_{\bar{p}} \cup \{M \cap H_{\mathfrak{c}^+}\}, \mathcal{N}_{\bar{p}} \cup \{M \cap H_{\mathfrak{c}^+}\})$$
will now do the trick (i.e., will be $(M, \mathcal{P})$-generic).

Recall that a set of reals X is $\aleph_1$-*dense* if $|X \cap (p, q)| = \aleph_1$ for all $p < q$ in $\mathbb{R}$.

Corollary 94. *Assuming* $\mathrm{mm} > \omega_1$, *every two* $\aleph_1$-*dense sets of reals are isomorphic.*

Proof. Fix two $\aleph_1$-dense sets of reals X and Y. Using Theorem 93, for every pair $I_{pq} = (p, q)$ and $I_{rs} = (r, s)$ of rational intervals we can fix increasing functions
$$f^{rs}_{pq} : I_{pq} \cap X \to I_{rs} \cap Y \text{ and } g^{rs}_{pq} : I_{pq} \cap Y \to I_{rs} \cap X.$$

Let $\mathcal{P}$ the poset of finite strictly increasing partial functions p from X into Y such that for every $x \in \mathrm{dom}(p)$ there exist a pair I_{pq} and I_{rs} of rational intervals such that either $p(x) = f^{rs}_{pq}(x)$ or $g^{rs}_{pq}(p(x)) = x$. We order $\mathcal{P}$ by the inclusion.

Claim 94.1. $\mathcal{P}$ *is a σ-centered poset.*

Proof. Fix a $p \in \mathcal{P}$. Let $|\mathrm{dom}(p)| = n$ and let x_i $(i < n)$ lists $\mathrm{dom}(p)$ according to the increasing ordering. Choose a sequence K_i $(i < n)$ of pairwise rational intervals that separates the x_i $(i < n)$ and another sequence L_i $(i < n)$ of pairwise disjoint rational intervals that separate the points of the sequence $p(x_i)$ $(i < n)$. For each $i < n$ we fix one of the functions of the form f_{pq}^{rs} or g_{pq}^{rs} such that either $p(x_i) = h_i(x_i)$ or $h_i(p(x)) = x$. Now note that if $p, q \in \mathcal{P}$ have all these invariants the same then $p \cup q$ belongs to $\mathcal{P}$. In fact, the same is true for every finite set of conditions from $\mathcal{P}$ that have common invariants. Since clearly there exist only countably many invariants, the posets $\mathcal{P}$ is σ-centered. $\square$

For $x \in X$ and $y \in Y$, set

$$\mathcal{D}_x = \{p \in \mathcal{P} : x \in \mathrm{dom}(p)\} \text{ and } \mathcal{D}^y = \{p \in \mathcal{P} : y \in \mathrm{range}(p)\}.$$

Then it is easily seen that all these sets are dense-open in $\mathcal{P}$, and that a filter that meets them all will give us the desired isomorphism between X and Y. $\square$

From now on, we fix a set of reals B of cardinality $\aleph_1$. It follows that, assuming $\mathfrak{mm} > \omega_1$, B will appear as a member of basis for the class of uncountable linear orderings, if such a basis exists at all. So, it remains to concentrate on the orthogonal to $\{\omega_1, \omega_1^*, B\}$. This class of uncountable linear orderings is the subject matter of our next definition.

Definition. A linearly ordered set L is *Aronszajn* if it is uncountable and if it contains no uncountable subordering which is well-ordered, conversely well-ordered, or separable. The name comes from the fact that any partition tree of such an ordering is Aronszajn and that a lexicographical ordering of any Aronszajn tree is an Aronszajn ordering (see, for example, [26]).

It remains to find a basis for the class of Aronszajn orderings. We shall do this in the next two sections of this Chapter.

15.3 Ordered coherent trees

The results of Chapter 9 suggest that coherent trees are likely to be relevant to questions surrounded the class of Aronszajn trees and therefore in many questions about uncountable structures. In this section we give one example of such a phenomenon by considering the basis problem for uncountable linear orderings. We start by analyzing the lexicographical orderings of coherent trees.

Theorem 95. *The cartesian square of a lexicographically ordered coherent tree that admits a strictly increasing mapping into the rationals can be covered by countably many chains.*

Proof. Let T be a given special coherent tree. Identifying a node $t : \alpha \to \omega$ of T with the characteristic function $\chi_{\Gamma(t)}$ of its graph $\Gamma(t) = \{(\xi, t(\xi)) : \xi < \alpha\}$ the natural bijection between $\alpha \times \omega$ and the ordinal $\omega\alpha$ will move $\chi_{\Gamma(t)}$ to a function $\bar{t} : \omega\alpha \to 2$. Since the transformation $t \mapsto \bar{t}$ preserves the lexicographical ordering and since the downward closure of $\{\bar{t} : t \in T\}$ remains special, we may assume that out tree T itself consists of maps whose ranges are included in $\{0, 1\}$. By our assumption, the tree T can be decomposed into countably many antichains. So let $a : T \to \omega$ be a fixed map such that $a(s) \neq a(t)$, whenever $s <_T t$ and such that a is one-to-one on the levels of T. It suffices to decompose the subset $\{(s, t) : \mathrm{ht}(s) \leq \mathrm{ht}(t)\}$ of the cartesian square of $(T, <_{\mathrm{lex}})$ into countably many chains. Note that the existence of a allows us to replace (modulo a countable decomposition) every such ordered pair (s, t) by the pair $(s, t \restriction \mathrm{ht}(s))$. It follows that it suffices to show that if $f : D \to T$ is a level-preserving map from a subset X of T which intersects a given level of T in at most one point, then its graph $\Gamma_f = \{(x, f(x)) : x \in X\}$ can be decomposed into countably many cartesian chains. For $t \in X$, let

$$D_t = \{\xi \in \omega_1 : \xi = \mathrm{ht}(t) \text{ or } \xi < \mathrm{ht}(t) \text{ and } t(\xi) \neq f(t)(\xi)\}.$$

Then by the coherence of the tree T, it follows that D_t is a finite set of countable ordinals for all $t \in X$. Let $p_t : D_t \to \omega$ and $q_t : D_t \to \omega$ be the functions defined by

$$p_t(\xi) = a(t \restriction \xi) \text{ and } q_t(\xi) = a(f(t) \restriction \xi).$$

It suffices to show that if for some $s <_{\mathrm{lex}} t$ in X, we have that

$$t \restriction D_t \cong s \restriction D_s, \quad p_t \restriction D_t \cong p_s \restriction D_s, \quad \text{and} \quad q_t \restriction D_t \cong q_s \restriction D_s,$$

then $f(s) <_{\mathrm{lex}} f(t)$. Note that from the isomorphisms between these parameters it follows in particular that s and t as well as $f(s)$ and $f(t)$ are incomparable in T.

Let us first establish the inequality

$$\Delta(f(s), f(t)) \geq \Delta(s, t).$$

To this end, let $\alpha = \Delta(s, t)$. Then $\alpha < \mathrm{ht}(s), \mathrm{ht}(t)$, since s and t are incomparable. By the properties of our parameters s and t, we have that $D_s \cap \alpha = D_t \cap \alpha$ and that $f(s)$ and $f(t)$ agree on this set. From the definition of D_s and D_t, we conclude that $f(s)$ and $f(t)$ must agree below α, and so $\Delta(f(s), f(t)) \geq \alpha$. Using the symmetry, we get that the other inequality

$$\Delta(s, t) \geq \Delta(f(s), f(t))$$

is true as well, and so we have that $\alpha = \Delta(s,t) = \Delta(f(s), f(t))$. Note that α cannot belong to the intersection of D_s and D_t since then it would occupy the same position in these sets, and so in particular we would have that $s(\alpha) = t(\alpha)$ contradicting the assumption that $\alpha = \Delta(s,t)$. So, we are left with the following three cases.

Suppose first that $\alpha \in D_s \setminus D_t$. Then

$$s(\alpha) < t(\alpha) = f(t)(\alpha),$$

and since $f(t)(\alpha)$ can take only two values 0 and 1, we conclude that $f(t)(\alpha) = 1$. Since $f(s)(\alpha) \neq f(t)(\alpha)$, using the same reasoning, we conclude that $f(s)(\alpha) = 0$, giving as the desired inequality $f(s) <_{\text{lex}} f(t)$.

Suppose now that $\alpha \in D_t \setminus D_s$. Then

$$f(s)(\alpha) = s(\alpha) < t(\alpha),$$

and so as before, we must have that $f(s)(\alpha) = 0$. Since $f(s)(\alpha) \neq f(t)(\alpha)$, we must have that $f(t)(\alpha) = 1$, giving as again the desired inequality $f(s) <_{\text{lex}} f(t)$.

It remains to consider the case $\alpha \notin D_t \cup D_s$. Then

$$f(s)(\alpha) = s(\alpha) < t(\alpha) = f(t)(\alpha)$$

giving us the desired inequality $f(s) <_{\text{lex}} f(t)$. $\qquad\square$

This result has the following consequence that will be useful for us later in this Chapter.

Theorem 96. *Assuming* $\mathfrak{m} > \omega_1$ *the cartesian square of lexicographically ordered coherent Aronszajn tree is the union of countably many chains.* $\square$

As indicated before there are many natural examples of coherent Aronszajn trees. For example the trees $T(\rho_i)$ for $i = 0, 1, 2, 3$ are such trees.

Theorem 97. *Assuming* $\mathfrak{mm} > \omega_1$, *any lexicographically ordered coherent Aronszajn tree is a* minimal *uncountable linear ordering in the sense that it embeds into all of its uncountable suborderings.*

Proof. Note that any coherent Aronszajn subtree T of $\omega^{<\omega_1}$ has the form

$$T(a) = \{a_\beta \restriction \alpha : \alpha \leq \beta < \omega_1\}$$

for some coherent mapping $a : [\omega_1]^2 \to \omega$. We shall prove the conclusion of the theorem not for the full $T(a)$ ordered lexicographically, but only for its subset $\{a_\beta : \beta < \omega_1\}$. In fact this suggests defining the ordering $<_a$ on ω_1 by letting

$$\alpha <_a \beta \text{ iff } a_\alpha <_{lex} a_\beta$$

and proving only that $(\omega_1, <_a)$ is a minimal uncountable linear ordering. This will bu sufficient for our needs below.

Let Ω be a given uncountable subset of ω_1. We need to construct a mapping $f : \omega_1 \to \Omega$ that is strictly increasing relative to the ordering $<_a$. This will be quite similar to the proof of Theorem 93 above. We do need however the following piece of notation: For a subset E of ω_1 and $\xi < \omega_1$, let $E(\xi)$ denote the maximal element of E that is smaller or equal to ξ ($E(\xi) = 0$ if such element does not exist). Thus, as before, we let $\mathcal{P}$ be the set of all triples $p = (f_p, \mathcal{M}_p, \mathcal{N}_p)$ such that

(1) f_p is a finite strictly $<_a$-increasing partial function from ω_1 into Ω.

(2) $\mathcal{N}_p$ is a finite $\in$-chain of countable elementary submodels of $H_{\aleph_2}$ that contain a and Ω.

(3) $\mathcal{M}_p \subseteq \mathcal{N}_p$ and for every $M \in \mathcal{M}_p$ there is a countable elementary submodel $\bar{M}$ of $H_{(2^{\aleph_1})^+}$ such that $M = \bar{M} \cap H_{\aleph_2}$.

(4) The range of f_p is separated by $\mathcal{N}_p$.

(5) For every $\xi \in \mathrm{dom}(f)$ there is $N \in \mathcal{N}_p$ such that $\xi \in N$ but $f_p(\xi) \notin N$ but there is no $M \in \mathcal{M}_p$ that separates ξ and $f_p(\xi)$.

Moreover, letting
$$D_p = \{N \cap \omega_1 : N \in \mathcal{N}_p\},$$
we have the following two additional conditions:

(6) $D_p(\Delta(\alpha, \beta)) = D_p(\Delta(f_p(\alpha), f_p(\beta)))$ for all $\alpha \neq \beta \in \mathrm{dom}(f_p)$,

(7) $D_p(\Delta(\alpha, \beta)) > D_p(\Delta(\beta, \gamma))$ iff $D_p(\Delta(f_p(\alpha), f_p(\beta))) > D_p(\Delta(f_p(\beta), f_p(\gamma)))$ for every triple $\alpha, \beta, \gamma \in \mathrm{dom}(f_p)$.[1]

The ordering on $\mathcal{P}$ are the set inclusions.

Claim 97.1. $\mathcal{P}$ is stationary set preserving.

Proof. Let $(C_p : p \in \mathcal{P})$ be a given $\mathcal{P}$-club. Let E be a stationary subset of ω_1, and $\bar{p} \in \mathcal{P}$. We need to find p extending $\bar{p}$ such that $C(p) \cap E \neq \emptyset$. Choose a countable elementary submodel M of $H_{(2^{\aleph_1})^+}$ containing all these objects such that $\delta = M \cap \omega_1 \in E$. Let

$$p = (f_{\bar{p}}, \mathcal{M}_{\bar{p}} \cup \{M \cap H_{\aleph_2}\}, \mathcal{N}_{\bar{p}} \cup \{M \cap H_{\aleph_2}\}).$$

Note that p belongs to $\mathcal{P}$. Choose q extending p such that $C(q) \setminus \delta \neq \emptyset$. It suffices to show that δ belongs to $C(q)$. Otherwise, we can find $r \leq q$ and $\gamma < \delta$ such that
$$(\forall s \leq r) \ C(s) \cap (\gamma, \delta] = \emptyset.$$

---

[1] When this happens, we say that f_p preserves *splitting patterns* relative to D_p.

Increasing γ if necessary, we assume that $\gamma > \max(D_r \cap \delta)$ and that

$$(\forall \xi, \eta \in \operatorname{dom}(f_r))[\Delta(\xi, \eta) < \delta \text{ implies } \Delta(\xi, \eta) < \gamma \text{ and } \Delta(f_r(\xi), f_r(\eta)) < \gamma].$$

Let $\mathcal{X}$ be the collection of all conditions x of $\mathcal{P}$ that end-extend $r \cap M$ and realize the same type as r over the parameters γ and $r \cap M$. This in particular means that we may assume that if $n = |f_r \setminus M|$ and if $\mathcal{C} \in M$ is a chain in the $2n$th cartesian power of $(\omega_1, <_a)$ such that $(\operatorname{dom}(f_r) \cup \operatorname{rang}(f_r)) \setminus M$ (considered as an $2n$-tuple ordered according to, say, $<_a$) belongs to $\mathcal{C}$ then this is also true for all other conditions $x \in \mathcal{X}$. Choose $x \in \mathcal{X} \cap M$ such that if $\delta_x \in D_x$ corresponds to $\delta \in D_r$, then the projections

$$\{a_\xi \upharpoonright \delta : \xi \in (\operatorname{dom}(f_r) \cup \operatorname{rang}(f_r)) \setminus \delta\} \text{ and}$$
$$\{a_\xi \upharpoonright \delta_x : \xi \in (\operatorname{dom}(f_x) \cup \operatorname{rang}(f_x)) \setminus \delta_x\}$$

are pairwise incomparable in the sense that every node of the first set is incomparable with every node of the second set. It follows that $f_r \cup f_x$ is a $<_a$-preserving partial map and that

$$s = (f_r \cup f_x, \mathcal{M}_r \cup \mathcal{M}_x, \mathcal{N}_r \cup \mathcal{N}_x)$$

satisfies the conditions (1)-(7) of being an element of our poset $\mathcal{P}$. However, note that $C(s)$ being a superset of $C(x)$ contains an ordinal from $[\gamma, \delta)$, a contradiction. This finishes the proof. $\square$

As before, using the way we set up the side conditions, one easily checks that for all $\alpha < \omega_1$, the set

$$\mathcal{D}_\alpha = \{p \in \mathcal{P} : \alpha \in \operatorname{dom}(f_p)\}$$

is a dense subset of $\mathcal{P}$. So an application of $\mathfrak{mm} > \omega_1$ gives us a mapping $f : \omega_1 \to \Omega$ that is strictly increasing relative to the ordering $<_a$. $\square$

From now one we fix a coherent mapping $a : [\omega_1]^2 \to \omega$ such that the corresponding tree $T(a)$ is special Aronszajn, and consider the corresponding linearly ordered set $C(a) = (\omega_1, <_a)$ for which we know that it has the property that its cartesian square is the union of countably many chains. We shall need the following corollary of the previous theorem.

Corollary 98. *Assuming* $\mathfrak{mm} > \omega_1$, *every uncountable linear ordering L whose cartesian square is the union of countably many chains contains an isomorphic copy of the ordering $C(a)$ or its reverse $C(a)^*$.*

Proof. Choose an arbitrary one-to-one mapping $f : C(a) \to L$. Let $\mathcal{P}$ be the set of all finite subsets of $C(a)$ on which f is strictly increasing. We consider $\mathcal{P}$ as a partially ordered set ordered by the inclusion. If $\mathcal{P}$ satisfies the countable chain condition then an application of $\mathfrak{m} > \omega_1$ would give us an uncountable subset K of $C(a)$ on which f is strictly increasing. It follows that $K \leq L$. Since by Theorem 97, $C(a) \leq K$ this gives us the conclusion of the Corollary. So we are left with analyzing the case when $\mathcal{P}$ fails to satisfy the countable chain condition. So let $\mathcal{X}$ be a given uncountable subset of $\mathcal{P}$ consisting of pairwise incompatible members of $\mathcal{P}$. Applying the Δ-system lemma and noting that the root can't contribute to the incompatibilities of members of $\mathcal{P}$, subtracting it from each member of $\mathcal{P}$, we may assume that $\mathcal{X}$ consists of pairwise disjoint finite subsets of $C(a)$ all of some fixed size $k \geq 1$. We show that in this case there is an uncountable subset K on which f is decreasing. Recall that $C(a)$ can naturally be viewed as a subset of the lexicographically ordered tree $T(a)$. Similarly, the ordering L can be identified with a subset of a lexicographically ordered Aronszajn tree T. So considering the $\wedge$-closures of elements of $\mathcal{X}$ as well as their f-images, we perform a Δ-system argument on them obtaining an uncountable subfamily $\mathcal{X}_0$ of $\mathcal{X}$ where the corresponding two families of trees form Δ-systems with roots R and S, respectively. Moreover, as in the previous proof we assume that the corresponding structures are isomorphic via natural isomorphisms. Note now that since for two different elements p and q of $\mathcal{X}_0$ the mapping f is not increasing on the union $p \cup q$, there must be $\alpha \in p$ and $\beta \in q$ occupying the same position in the increasing enumerations of p and q, respectively such that

$$\alpha <_a \beta \text{ iff } f(\alpha) >_a f(\beta). \tag{15.1}$$

As above, we may identify a $p \in \mathcal{X}_0$ with the k-tuple $\langle p(1), ..., p(k) \rangle$ of elements of $C(a)$ that enumerates p increasingly according to the ordering of $C(a)$. Similarly, we identify the image $f"p$ with the k-tuple $\langle f(p(1)), ..., f(p(k)) \rangle$ of elements of L. So going to an uncountable subfamily of $\mathcal{X}_0$, we may assume that $\{\langle p(1), ..., p(k) \rangle : p \in \mathcal{X}_0\}$ is a chain in $C(a)^k$ and that $\{\langle f(p(1)), ..., f(p(k)) \rangle : p \in \mathcal{X}_0\}$ is a chain in L^k. Using (15.1) it follows now that f is decreasing on $K = \{p(1) : p \in \mathcal{X}_0\}$. Combining this with Theorem 97, we get the desired conclusion $C(a)^* \leq K^* \leq L$ of the Corollary.   □

15.4 Aronszajn orderings

In this section we prove the following result which shows that uncountable linear orderings have a two-element basis.

Theorem 99. *Assuming* mm $> \omega_1$, *every Aronszajn ordering contains an uncountable subordering whose cartesian square can be covered by countably many chains.*

Proof. So let L be a given Aronszajn ordering and let T be one of its binary partition trees (see [26]). Thus L can naturally be identified with the subset $\{\{x\} : x \in L\}$ of T. Choose an uncountable subset X of $T(\rho_3)$

which intersects a given level of T in at most one point and a level-preserving map

$$f : X \to L \subseteq T.$$

By $\mathfrak{mm} > \omega_1$, going to an uncountable subset of X, we may assume that f preserves splitting patterns, i.e., that for all $x, y, z \in \mathrm{dom}(f)$,

$$\triangle(x, y) > \triangle(x, z) \text{ iff } \triangle(f(x), f(y)) > \triangle(f(x), f(z)). \qquad (15.2)$$

To see this, let $\mathcal{P}$ be the set of all pairs $p = (X_p, D_p)$ of finite subsets of $\mathrm{dom}(f)$ and ω_1, respectively, such that (15.2) holds for all $x, y, z \in X_p$ and such that for all $x, y \in X_p$ and $\xi \in D_p$,

$$\triangle(x, y) < \xi \text{ iff } \triangle(f(x), f(y)) < \xi. \qquad (15.3)$$

We let $\mathcal{P}$ be ordered by coordinatewise inclusions. To see that $\mathcal{P}$ is stationary-set preserving pick a $\mathcal{P}$-club $(C(p) : p \in \mathcal{P})$, $\bar{p} \in \mathcal{P}$ and a stationary subset E of ω_1. We need to find an extension p of $\bar{p}$ such that $C(p) \cap E \neq \emptyset$. To see this, choose M, a countable elementary submodel of some large-enough structure of the form $(H_\theta, \in)$, such that $\delta = M \cap \omega_1 \in E$. Let $p = (X_{\bar{p}}, D_{\bar{p}} \cup \{M \cap \omega_1\})$. Then p is an M-generic condition of $\mathcal{P}$, so $\delta \in C(q)$ for any extension q of p with the property that $C(q) \setminus \delta \neq \emptyset$. Note also that a condition of the form $(\{x\}, \{\xi\})$, where $\xi = M \cap \omega_1$ and $x \in \mathrm{dom}(f) \setminus M$ for such a model M forces that the unions of the first projection of the generic filter is uncountable. Hence an application of the $\mathfrak{mm} > \omega_1$ to $\mathcal{P}$ restricted below this condition gives us the conclusion we need, an uncountable subset of the domain of f for which the condition (15.2) holds.

Applying Theorem 67 and going to an uncountable subset of X, we may assume that f is Lipschitz (the case that f^{-1} is Lipschitz is considered similarly). Let $U(f)$ be the downwards closure of the graph of f in the tree-product $T(\rho_3) \otimes T$. Shrinking X if necessary we may assume that $U(f)$ is also a binary tree. This is so because the poset of finite subsets of the graph of f whose downward closures are binary satisfies the countable chain condition. Now, chose an uncountable subset Y of X and an unbounded subset D of ω_1, such that

$$[\triangle(x, y), \triangle(f(x), f(y))] \cap D = \emptyset \text{ for all } x \neq y \in Y. \qquad (15.4)$$

That this can be done follows from the fact that the natural poset of finite approximations to D and T is easily seen to satisfy the countable chain condition using the fact that f preserves splitting patterns, i.e., satisfies the condition (15.2). Taking the closure of D we may assume that D is closed and unbounded. Going to an uncountable subset of Y, we may assume that the heights of every pair of comparable splitting nodes of $U(f \restriction Y)$ are separated by a member of D. That this can be done follows from the fact that the poset of finite subsets of Y with this property satisfies the countable chain condition. It follows that to every splitting node $s = x \wedge y$ of the subtree $S(Y)$ of $T(\rho_3)$ generated by taking the downwards closure of Y there corresponds a splitting node $t_s = f(x) \wedge f(y)$ and the association $s \mapsto t$ does not depend on which x and y we pick in Y to represent s. Let K be the collection of all splitting nodes $s = x \wedge y$ of $S(Y)$ such that $x <_{\text{lex}} y$ iff $f(x) <_L f(y)$. Note again that the definition does not depend on the x and y in Y we choose to represent s since the reason for the inequality $f(x) <_L f(y)$ rely on the fact that $f(x)$ must lie in the left and $f(y)$ in the right immediate successor of the splitting node $t_s = f(x) \wedge f(y$. By Theorem 67, there is an uncountable subset Z of Y such that either

(1) $\{x \wedge y : x, y \in Z, x \neq y\} \subseteq K$, or

(2) $\{x \wedge y : x, y \in Z, x \neq y\} \cap K = \emptyset$.

In the first case, the restriction $f : (Z, <_{\text{lex}}) \to L$ is strictly increasing and in the second case, it is strictly decreasing. In any case, by Corollary 96, the image $f"Z$ is an uncountable subordering of L whose cartesian square can be covered by countably many chains. $\square$

Combining Theorems 93 and 99 and Corollary 98, we obtain the following.

Theorem 100. *Assuming* $\mathfrak{mm} > \omega_1$, *a linearly ordered set L is countable if and only if it contains no isomorphic copy of any ordering from the list* ω_1, ω_1^*, B, $C(a)$, *and* $C(a)^*$. $\square$

As indicated above the role of B in this result can be played by any set of reals of size $\aleph_1$ and the role of $C(a)$ can be played by any uncountable linear ordering C whose cartesian square is the union of countably many chains.

Chapter 16

Cardinal Arithmetic and mm

16.1   mm and the continuum

Theorem 101. mm $> \omega_1$ *implies* mm $= \mathfrak{c} = \omega_2$.

Proof. Assumig mm $> \omega_1$ we will construct a one to one map from ω^ω into ω_2. We procceed as follows. Fix a family $\{S_n : n \in \mathbb{N}\} \subseteq stat(\omega_1)$ and $\{T_n : n \in \mathbb{N}\} \subseteq stat\{\alpha < \omega_2 : cf(\alpha) = \omega\}$ such that

1. $\bigcup\limits_{n=0}^{\infty} S_n = \omega_1$,

2. $\bigcup\limits_{n=0}^{\infty} T_n = \{\alpha < \omega_2 : cf(\alpha) = \omega\}$,

3. $S_m \cap S_n = \emptyset$ and $T_m \cap T_n = \emptyset$ for each $m \neq n$.

For each $r \in \omega^\omega$ we define an ordinal $\delta(r)$ (we will use it to define the desired one to one function from ω^ω into ω_2),

$\delta(r) = min\{\delta : cof(\delta) = \omega_1, (\exists C \subseteq \delta\ club)(otp(C) = \omega_1)(\forall\gamma \in C)(\forall n \in \omega)(otp(C \cap \gamma) \in S_n \leftrightarrow \gamma \in T_{r(n)})\}$.

Claim 101.1. *If* mm $> \omega_1$, *then for every* $r \in \omega^\omega$, *the ordinal* $\delta(r)$ *exists and* $\delta(r) < \omega_2$.

Proof. Fix $r \in \omega^\omega$. Consider the following partial order

$\mathcal{P} = \{p : p \subseteq \omega_2, (|p| \leq \omega)(\gamma \in p)(\forall n \in \omega)(otp(p \cap \gamma) \in S_n \leftrightarrow \gamma \in T_{r(n)})\}$.

We order $\mathcal{P}$ by end extension. In order to apply that $\mathrm{mm} > \omega_1$ we need to show the following

Claim 101.2. $\mathcal{P}$ *is stationary set preserving.*

Proof. Let $\langle C(p) : p \in \mathcal{P} \rangle$ a given $\mathcal{P}$-club and fix $E \in stat(\omega_1)$. Let $\bar{p} \in \mathcal{P}$ arbitrary. We need to find $p \leq \bar{p}$ such that $C(p) \cap E \neq \emptyset$.

Fix $n \in \mathbb{N}$ such that $E \cap S_n \in Stat(\omega_1)$. Find an elementary submodel M of $(H_{\mathfrak{c}^+}, \in)$ of cardinality $\aleph_1$ such that $\gamma = M \cap \omega_2 \in T_{r(n)}$, and such that r, $\langle C(p) : p \in \mathcal{P} \rangle$, E, $\bar{p}$ belong to M. Note that $M \in H_{\mathfrak{c}^+}$. Let N be a countable elementary submodel N of $(H_{\mathfrak{c}^+}, \in)$ containig r, $\langle C(p) : p \in \mathcal{P} \rangle$, E, $\bar{p}$ and containing also M. We also require that $\alpha = N \cap \omega_1 \in E \cap S_n$. Let $N^* = M \cap N$. Note that in this case N^* is a countable elementary submodel of $(H_{\mathfrak{c}^+}, \in)$ and $\alpha = N^* \cap \omega_1$.

Let $\gamma = sup(N^* \cap \omega_2)$ take $\{\mathcal{X}_n : n \in \omega\}$ an enumeration of all subsets of $\mathcal{P}$ which are elements of N^*. Starting from $p_0 = \bar{p}$, we build as before a sequence $p_0 \geq p_1 \geq p_2 \ldots$ of elements of $N^* \cap \mathcal{P}$ such that for all n either $p_{n+1} \in \mathcal{X}_n$ or there is no $q \leq p_n$ such that $q \in \mathcal{X}_n$. We construct the sequence inductively, suppose we have $\{p_i : i \leq n\}$ already defined. Let $\gamma' \in N^* \cap \gamma$ such that $\mathcal{X}_n = \{p \in \mathcal{P} : max(p) \geq \gamma'\}$. In order to define p_{n+1}, let $\gamma_n = max(p_n)$. Choose $k \in \mathbb{N}$ such that $otp(p_n \cap \gamma_n) + 1 \in S_k \leftrightarrow \gamma_n \in T_{r(k)}$. Since $T_{r(k)}$ is unbounded and by elementarity, we can take $\gamma^* \geq \gamma'$ in $T_{r(k)} \cap N^*$. Define $p_{n+1} = p_n \cup \{\gamma^*\}$.

$$ p_\omega = (\bigcup_{n=0}^{\infty} p_n) \cup \{\gamma\}. $$

Given $\gamma' < \gamma$, we have that $\{p \in \mathcal{P} : otp(p) > \gamma'\}$ is an element of N^*, i.e., it is one of the $\mathcal{X}_n$. In order to prove that $sup(\bigcup p_n) = \gamma$ it is enough to show that for every $\gamma' \in N^* \cap \gamma$ and every n such that $\mathcal{X}_n = \{p \in \mathcal{P} : otp(p) > \gamma'\}$ we have that $p_{n+1} \in \mathcal{X}_n$, which is true by the way we constructed our sequence p_n.

Now, find $p \leq p_\omega$ such that $C(p) \setminus \alpha \neq \emptyset$. We claim that $\alpha \in C(p) \cap E$. Suppose it is not the case, then by definition of $\mathcal{P}$-club there is $q \leq p$ such that $C(r) \cap [\beta, \alpha) = \emptyset$ for every $r \leq q$. Let $\mathcal{X} = \{r \in \mathcal{P} : C(r) \setminus \beta \neq \emptyset\}$. Then $\mathcal{X} \in N^*$ and $q \in \mathcal{X}$. Pick n such that $\mathcal{X} = \mathcal{X}_n$. Since $q \in \mathcal{X}_n$ and $q \leq p$, we must have that $p_{n+1} \in \mathcal{X}_n$. Using tha fact that $p_{n+1} \in N^*$ we have that $C(p_{n+1}) \cap [\beta, \alpha) \neq \emptyset$. By monotonicity, $q \leq p_{n+1}$ implies that $C(q) \cap [\beta, \alpha) \neq \emptyset$, wich is a contradiction. Hence $\alpha \in E \cap C(p)$. so, we conclude that $\mathcal{P}$ is stationary set preserving.

For $\alpha < \omega_1$, define

$$\mathcal{D}_\alpha = \{p \in \mathcal{P} : otp(p) \geq \alpha\}$$

We prove that $\mathcal{D}_\alpha$ is dense open for each $\alpha < \omega_1$. Let $\alpha*$ be a limit ordinal. Pick $\bar{p} \in \mathcal{P}$ such that no $p \leq \bar{p}$ belongs to $\mathcal{D}_\alpha$. Find a countable elementary submodel M' of $(H_{\mathfrak{c}+}, \epsilon)$ containing $r, \bar{p}$, E and $\rangle C(p) : p \in \mathcal{P}$ and such that $\alpha = M \cap \omega_1 \in S_0$. Then $\gamma = sup(M \cap \omega_2) \in T_{r(0)}$. Let $\{\mathcal{X}_n : n \in \omega\}$ be an enumeration of all the subsets odf $\mathcal{P}$ which are elements of M' and construct a sequence $\bar{p} = p_0 \geq p_1 \geq \ldots$ of elements of $\mathcal{P} \cap M$ such that $p_{n+1} \in \mathcal{X}_n$ or there is no $q \leq p_n$ such that $q \in \mathcal{X}_n$ (we do it as before). Define $p_\omega = (\bigcup\limits_{n=0}^{\infty} p_n) \cap \gamma$. Then $p_\omega \in \mathcal{D}_\beta$ for each $\beta < \alpha^*$. Since $\{\mathcal{D}_\beta : \beta < \alpha^*\} \subseteq \{\mathcal{X}_n : n \in \omega\}$, we must have that $otp(p_\omega) \geq \beta$, for every $\beta < \alpha^*$, since α^* is a limit ordinal, we have that $otp(p_\omega) \geq \alpha^*$ as desired. $\square$

Applying mm $> \omega_1$ we can find $\mathcal{G}$ a generic filter intersecting $\mathcal{D}_\alpha$ for each $\alpha < \omega_1$. Define $C = \cup\mathcal{G}$ and $\delta = supC$. Then C is a club subset of δ of order type ω_1. We also have that

$$(\forall \gamma \in C)(\forall n \in \omega)(otp(C \cap \gamma \in S_n \leftrightarrow \gamma \in T_{r(n)})). \text{ It follows that } \delta(r) \leq \delta.$$

$\square$

Note that $\delta(r) \neq \delta(r')$ whenever $r \neq r'$. If $r \neq r'$, find $n \in \omega$ such that $r(n) \neq r'(n)$ and assume that $\delta(r) = \delta(r') = \delta$. Let C and C' be te correspondent clubs in δ witnessing that $\delta(r) = \delta(r') = \delta$. Define

$$D = \{\gamma \in C \cap C' : otp(C \cap \gamma) = otp(C' \cap \gamma)\}.$$

Note that D is club in γ. Define

$$\bar{D} = \{otp(C \cap \gamma) : \gamma \in D\}.$$

Then $\bar{D}$ is a club in ω_1, therefore $\bar{D} \cap S_n \neq \emptyset$. So we can choose $\gamma \in D$ such that $otp(C \cap \gamma) = otp(C' \cap \gamma) \in S_n$. Since C witness that $\delta(r) = \delta$, we conclude that $\gamma \in T_{r(n)}$, similarly we conclude that $\gamma \in T_{r'(n)}$, wich is a contradiction to the fact that $T_{r(n)} \cap T_{r'(n)} = \emptyset$. Hence we can define a one to one function from ω^ω into ω_2 (we assign to each $r \in \omega^\omega$ the ordinal $\delta(r)$).

Since mm $\leq \mathfrak{c}$, we conclude that mm $> \omega_1$ implies mm $= \mathfrak{c} = \omega_2$. $\square$

16.2   mm and cardinal arithmetic above the continuum

In this section we prove the following fact.

Theorem 102. mm $> \omega_1$ *implies that* $\theta^{\aleph_1} = \theta$ *for all regular* $\theta \geq \omega_2$.

Proof. This is really just a variation on the previous proof so we only give a sketch leaving the rest of the details to the interested reader. We start again with two families $\{S_\xi : \xi < \omega_1\} \subseteq stat(\omega_1)$ and $\{T_\eta : \eta < \theta\} \subseteq stat\{\alpha < \theta : cf(\alpha) = \omega\}$ such that

1. $\triangle_{\xi < \omega_1} S_\xi = \omega_1$,

2. $\bigcup_{\eta < \theta} T_\eta = \{\alpha < \theta : cf(\alpha) = \omega\}$,

3. $S_\xi \cap S_\eta = \emptyset$ and $T_\xi \cap T_\eta = \emptyset$ whenever $\xi \neq \eta$.

For each $r \in \theta^{\omega_1}$ we define an ordinal $\delta(r)$ which will give us the desired one to one function from θ^{ω_1} into θ,

$$\delta(r) = min\{\delta : cof(\delta) = \omega_1, (\exists C \subseteq \delta \ club)(otp(C) = \omega_1)(\forall \gamma \in C)(\forall \xi < \omega_1)(otp(C \cap \gamma) \in S_\xi \leftrightarrow \gamma \in T_{r(\xi)})\}.$$

As before, we arrive at the crucial fact.

Claim 102.1. *If* mm $> \omega_1$, *then for every* $r \in \theta^{\omega_1}$, *the* $\delta(r) < \theta$ *exists.*

Proof. Fix $r \in \theta^{\omega_1}$. Consider the following partial order

$$\mathcal{P} = \{p : p \subseteq \omega_2, (|p| \leq \omega)(\gamma \in p)(\forall \xi < \omega_1)(otp(p \cap \gamma) \in S_\xi \leftrightarrow \gamma \in T_{r(\xi)})\}.$$

We order $\mathcal{P}$ by end extension. The proof that $\mathcal{P}$ is stationary-set preserving is left to the reader as it is almost identical to the corresponding proof above.

□

□

Chapter 17

Reflection Principles

17.1 Strong reflection of stationary sets

Definition. The *Strong Reflection Principle* at a cardinal λ, denoted by SRP(λ), is the statement that for every $S \subseteq [\lambda]^\omega$ and $a \in H_{\lambda^+}$ there is a continuous $\in$-chain $N_\xi (\xi < \omega_1)$ of countable elementary submodels of $(H_{\lambda^+}, \in)$, all containing the prescribed set a, such that for all $\xi < \omega_1$, we have that $N_\xi \cap \lambda \in S$ if and only if we can find a countable elementary submodel M of $(H_{\lambda^+}, \in)$ such that $M \supseteq N_\xi$, $M \cap \omega_1 = N_\xi \cap \omega_1$, and $M \cap \lambda \in S$. Let the SRP be the assertion that SRP(λ) holds for all cardinals λ.

Theorem 103. $\mathfrak{mm} > \omega_1$ *implies* SRP.

Proof. Given $S \subseteq [\lambda]^\omega$ and $a \in H_{\lambda^+}$, let $\mathcal{P}$ be the posets of all countable $\in$-chains $p = (N_\xi : \xi \le \alpha_p)$ of countable elementary submodels of $(H_{\lambda^+}, \in)$, all containing the prescribed set a, such that for all $\xi \le \alpha_p$, we have that $N_\xi \cap \lambda \in S$ if and only if there is a countable elementary submodel M of $(H_{\lambda^+}, \in)$ such that $M \supseteq N_\xi$, $M \cap \omega_1 = N_\xi \cap \omega_1$, and $M \cap \lambda \in S$. We order $\mathcal{P}$ by end-extension.

Claim 103.1. $\mathcal{P}$ *is stationary set preserving and, in fact, semi-proper.*

Proof. Let $\langle C(p) : p \in \mathcal{P} \rangle$ a given $\mathcal{P}$-club and fix $E \in stat(\omega_1)$. Let $\bar{p} \in \mathcal{P}$ arbitrary. We need to find $p \le \bar{p}$ such that $C(p) \cap E \ne \emptyset$. Choose a countable elementary submodel M of $(H_{(2^\lambda)^+}, \in, <_w)$ containing all these objects such that $M \cap \omega_1 \in E$. Let $N = M \cap H_{\lambda^+}$. If there is countable $N^* \prec H_{\lambda^+}$ such that $N^* \supseteq N$, $N^* \cap \omega_1 = N \cap \omega_1$, and $N^* \cap \lambda \in S$ fix one and let

$$M^* = \mathrm{Sk}_{(H_{(2^\lambda)^+}, \in, <_w)} M \cup (N^* \cap \lambda).$$

193

Then M^* is a countable elementary submodel of $(H_{(2^\lambda)^+}, \in, <_w)$ containing all the relevant objects such that

$$M^* \cap \lambda = N^* \cap \lambda,$$

and so in particular, $M^* \cap \lambda \in S$ and $M^* \cap \omega_1 \in E$. If such an $N^* \prec H_{\lambda^+}$ cannot be found, let $M^* = M$. Let $\mathcal{D}_n$ $(n < \omega)$ be the enumeration of all dense-open subsets of $\mathcal{P}$ that belong to M^*. A simple recursion will give us a decreasing sequence p_n $(n < \omega)$ of elements of $\mathcal{P} \cap M^*$ such that $p_0 = \bar{p}$ and $p_{n+1} \in \mathcal{D}_n$ for all $n < \omega$. Let $\delta = \sup_{n<\omega} \alpha_{p_n}$ and let

$$p_\omega = (\bigcup_{n<\omega} p_n) \cup \{(\delta, M^* \cap \lambda)\}.$$

In fact, it is not difficult to see that $\delta = M^* \cap \omega_1$. Choose $p \leq p_\omega$ such that $C(p) \setminus \delta \neq \emptyset$. Working as in the corresponding part of the proof of Claim 101.2 above, we show that $\delta \in C(p)$, and therefore that $C(p) \cap E \neq \emptyset$. □

Note that this proof also shows that for every $\alpha < \omega_1$, the set

$$\mathcal{D}_\alpha = \{p \in \mathcal{P} : \alpha_p \geq \alpha\}$$

is dense open in $\mathcal{P}$, so a filter that intersects all these sets will give us a continuous $\in$-chain of length ω_1 sarisfying the conclusion of SRP for the given $S \subseteq [\lambda]^\omega$ and $a \in H_{\lambda^+}$. □

The Strong Reflection Principle has the following important consequence.

Theorem 104. SRP(ω_2) *implies* NS$_{\omega_1}$ *is saturated.*

Proof. Consider a sequence E_ξ $(\xi < \omega_2)$ of subsets of ω_1. Let

$$S = \{A \in [\omega_2]^\omega : (\exists \xi \in A) A \cap \omega_1 \in E_\xi\}.$$

Let $N_\alpha (\alpha < \omega_1)$ be a continuous $\in$-chain of countable elementary submodels of H_{ω_3}, all containing the sequence $(E_\xi : \xi < \omega_1)$, which strongly reflects the set S. Let $N_{\omega_1} = \bigcup_{\alpha<\omega_1} N_\alpha$. The following fact shows that $(E_\xi : \xi < \omega_1)$ could not have been an antichain in $\mathcal{P}(\omega_1)/\text{NS}_{\omega_1}$ finishing the proof.

Claim 104.1. *For all* $\xi \notin N_{\omega_1}$, *either* E_ξ *is not stationary, or there is* $\eta \in N_{\omega_1} \cap \omega_2$ *such that* $E_\xi \cap E_\eta$ *is stationary.*

Proof. If E_ξ is stationary, we can choose a countably elementary submodel M of H_{ω_3} containing all these objects such that

$$\xi \in M \text{ and } M \cap \omega_1 \in E_\xi.$$

Let $\delta = M \cap \omega_1$. Then from the assumption $(N_\alpha : \alpha < \omega_1) \in M$, we conclude that

$$N_\delta \subseteq M \text{ and } N_\delta \cap \omega_1 = M \cap \omega_1 = \delta.$$

Since, clearly $M \cap \omega_2 \in S$, we conclude that also $N_\delta \cap \omega_2 \in S$. So we can pick $\eta \in N_\delta \cap \omega_1$. It follows that $\xi, \eta \in M$ and $M \cap \omega_1 = \delta \in E_\xi \cap E_\eta$. So, by the elementarity of M, the intersection $E_\xi \cap E_\eta$ must be stationary, as required. □

This finishes the proof. □

Definition. A subset S of $[\lambda]^\omega$ where λ is an ordinal $\geq \omega_1$ is *projectively stationary* if for every stationary subset E of ω_1, the set

$$S(E) = \{A \in S : A \cap \omega_1 \in E\}$$

is stationary in $[\lambda]^\omega$.

Theorem 105. *Assume* SRP(λ) *for some cardinal* $\lambda \geq \omega_2$. *Then for every projectively stationary set* $S \subseteq [\lambda]^\omega$ *and for every set* $X \subseteq \lambda$ *such that* $\omega_1 \subseteq X$ *and* $|X| = \aleph_1$ *there is* $X \subseteq T \subseteq S$ *of cardinality* $\aleph_1$ *such that* $S \cap [T]^\omega$ *contains a closed and unbounded subset of* $[T]^\omega$.

Proof. Let $N_\xi (\xi < \omega_1)$ be a continuous $\in$-chain of countable elementary submodels of H_{λ^+}, all containing the given set X, that strongly reflects the set S. Let

$$T = \bigcup_{\xi < \omega_1} N_\xi \cap \lambda.$$

Let $E = \{\xi < \omega_1 : N_\xi \cap \lambda \notin S.\}$ Then from our assumption that S is projectively stationary, we easily conclude that E cannot be stationary. This concludes our proof. □

Remark. It turns out that the conclusion of Theorem 105 gives us another formulation of the Strong Reflection Principles which some people find more elegant.

17.2 Weak reflection of stationary sets

Definition. The *Weak Reflection Principle* at a cardinal $\lambda \geq \omega_2$, denoted by WRP(λ), is the statement that for every stationary set $S \subseteq [\lambda]^\omega$ than every set $X \subseteq \lambda$ such that $\omega_1 \subseteq X$ and $|X| = \aleph_1$ there is $X \subseteq T \subseteq S$ of cardinality $\aleph_1$ such that $S \cap [T]^\omega$ is a stationary subset of $[T]^\omega$.

Theorem 106. SRP(λ) *implies* WRP(λ).

Proof. Let $N_\xi(\xi < \omega_1)$ be a continuous $\in$-chain of countable elementary submodels of H_{λ^+}, all containing the given set X, that strongly reflects the stationary set S. Let

$$T = \bigcup_{\xi < \omega_1} N_\xi \cap \lambda \text{ and } E = \{\xi < \omega_1 : N_\xi \cap \lambda \in S\}.$$

Then, since S is stationary in $[\lambda]^\omega$, one easily conludes that the set E must also be stationary. This completes the proof. $\qquad\square$

The Weak Reflection Principle has an important consequence subject of our next definition.

Definition. The *Strong Chang Conjecture*, CC*, is the statement that for every countable elementary submodel N of some large-enough structure of the form $(H_\theta, \in, <_w)$ there is countable elementary submodel M of $(H_\theta, \in, <_w)$ such that $N \subseteq M$, $M \cap \omega_1 = M \cap \omega_1$ and $M \cap \omega_2 \neq N \cap \omega_2$.

Here is a typical application of this principle which shows its relevance to the Weak Reflection Principle.

Theorem 107. *CC* implies* WRP(ω_2) *which in turn implies* $2^{\aleph_0} \leq \aleph_2$.

Theorem 108. WRP(λ) *for* $\lambda = 2^{\aleph_2}$ *implies* CC*.

Proof. Let $I = \omega_1 \cup \omega_1^{\omega_2}$. For $A \in [I]^\omega$ let

$$D_A = \{\delta < \omega_1 : \forall f \in I \cap A \ f(\delta) \in A \cap \omega_1\}$$

and let

$$S = \{A \in [I]^\omega : D_A \text{ is bounded in } \omega_2\}.$$

Claim 108.1. WRP(λ) *implies that S is not stationary in* $[I]^\omega$.

Proof. Otherwise, we can find $\omega_1 \subseteq J \subseteq I$ such that $|J| = \aleph_1$ and such that $S \cap [J]^\omega$ is stationary in $[J]^\omega$. Fix a continuous increasing sequence J_ξ $(xi < \omega_1)$ of countable subsets of J that cover J such that

$$\alpha_\xi = J_\xi \cap \omega_1 \in \omega_1 \text{ for all } \xi < \omega_1.$$

Let $E = \{\xi < \omega_1 : J_\xi \in S\}$. Then E is a stationary subset of $omega_1$, For $\xi \in E$, let $\delta_\xi = \sup(D_{J_\xi})$ and let $\bar\delta = \sup_{\xi \in E} \delta_\xi$. Choose a countable elementary submodel M of H_{λ^+} containig all these objects such that $\eta = M \cap \omega_1$ belongs to E. Then $M \cap J = J_\eta$. Since $\bar\delta$ belongs to M as well we can choose $\delta \in M$ such that $\delta > \bar\delta \geq \delta_\eta$. However, by the elementarity of M for all $f \in J_\eta \cap I \subseteq M$ we have that $f(\delta) \in M \cap \omega_1 = \eta = \alpha_\eta = J_\eta \cap \omega_1$, contradicting the fact that D_{J_η} was bounded by $\bar\delta$. $\qquad\square$

Pick a countable elementary submodel N of some large-enough structure of the form $(H_\theta, \in, <_w)$. Then by the Claim and the elementarity of N it must be that $M \cap I \notin S$. This means that $D_{N \cap I}$ is unbounded in ω_2. Choose $\delta \in D_{N \cap I}$ such that $\delta > \sup(M \cap \omega_2)$. Let

$$M = \mathrm{Sk}_{(H_\theta, \in, <_w)}((N \cap I) \cup \{\delta\}).$$

It remains only to show that $M \cap \omega_1 = N \cap \omega_1$. So choose $\alpha \in M \cap \omega_1$. Then there is a Skolem function $f \in N$ such that $f(\delta) = \alpha$. We may assume that $f : \omega_2 \to \omega_1$. So, in particular, $f \in M \cap I$. But δ was chosen from $D_{N \cap I}$, so we conclude that $\alpha = f(\delta) \in N \cap \omega_1$, as required.

This finishes the proof.   □

17.3   Open stationary set-mapping reflection

Definition. Let λ be an uncountable cardinal and let $\theta = (\lambda^{\aleph_0})^+$. Let M be a countable elementary submodel of H_θ such that $\lambda \in M$. A set $S \subseteq [\lambda]^\omega$ is M-*stationary* if $S \cap C \neq \emptyset$ for all $C \in M$ that are closed and unbounded in $[\lambda]^\omega$.

The set $[\lambda]^\omega$ carries a natural generated by the collection of basic-open sets of the form

$$[a, A] = \{B \in [\lambda]^\omega : a \subseteq B \subseteq A\},$$

where $A \in [\lambda]^\omega$ and $a \in [A]^{<\omega}$. We shall only consider this topology on $[\lambda]^\omega$ and call a mapping with range power-set of $[\lambda]^\omega$ *open* whenever all of its values are open subsets of $[\lambda]^\omega$.

Definition. The *Open Stationary Set-Mapping Reflection*, denoted by OSMR or MRP, is the principle which asserts that for λ uncountable and $\theta = (\lambda^{\aleph_0})^+$, if

$$F : [H_\theta]^\omega \to \mathcal{P}([\lambda]^\omega)$$

is an *open* and *stationary* set-mapping, i.e., for each $M \in [H_\theta]^\omega$ the subset $F(M)$ of $[\lambda]^\omega$ is open and M-stationary, then for every $a \in H_\theta$ there is a continuous $\in$-chain N_ξ ($\xi < \omega_1$) of countable elementary submodels of the structure $(H_\theta, \in, <_w)$, all containing the given set a, such that for all limit $0 < \eta < \omega_1$ there is $\eta_0 < \eta$ such that $N_\xi \cap \lambda \in F(N_\eta)$ for all $\xi \in (\eta_0, \eta)$.

Theorem 109. $\mathfrak{mm} > \omega_1$ *implies the Open Stationary Set-Mapping Reflection.*

Proof. Let $\mathcal{P}$ consists of all countable continuous $\in$-chains $p = (N_\xi^p : \xi \leq \alpha_p)$ of countable elementary submodels of $(H_\theta, \in, <_w)$, all containing the given parameter $a \in H_\theta$, such that for all limit $0 < \eta \leq \alpha_p$ there is $\eta_0 < \eta$ such that $N_\xi^p \cap \lambda \in F(N_\eta^p)$ for all $\xi \in (\eta_0, \eta)$. We order $\mathcal{P}$ by end-extension.

Claim 109.1. $\mathcal{P}$ *is stationary set preserving.*

Proof. Let $\langle C(p) : p \in \mathcal{P} \rangle$ a given $\mathcal{P}$-club and fix $E \in stat(\omega_1)$. Let $\bar{p} \in \mathcal{P}$ arbitrary. We need to find $p \leq \bar{p}$ such that $C(p) \cap E \neq \emptyset$. Choose a countable elementary submodel M of some much larger structure of the form $(H_\kappa, \in)$ such that M contains $(H_{(2^\theta)^+}, \in, <_w)$ and all the above objects and such that $\delta = M \cap \omega_1 \in E$. Let $\mathcal{D}_n$ $(n < \omega)$ be an enumeration of all dense-open subsets of $\mathcal{P}$ that belong to M. We build recursively a decreasing sequence $\bar{p} = p_0 \geq p_1 \geq \cdots \geq p_n \geq \cdots$ of elements of $\mathcal{P} \cap M$ such that $p_{n+1} \in \mathcal{D}_n$ for all n. Suppose p_n is given. Let

$$ C_n = \{ A \in [\lambda]^\omega : A = \mathrm{Sk}_{(H_{(2^\theta)^+}, \in, <_w)}(A \cup \{p_n, \mathcal{D}_n\}) \cap \lambda \}. $$

Then C_n is a closed and unbounded subset of $[\lambda]^\omega$ and $C_n \in M \cap H_\theta$. For $A \in C_n$, let $A^* = \mathrm{Sk}_{(H_{(2^\theta)^+}, \in, <_w)}(A \cup \{p_n, \mathcal{D}_n\})$. Since $F(M \cap H_\theta)$ is $M \cap H_\theta$-stationary, we can find A_n in the intersection $C_n \cap M \cap F(M \cap H_\theta)$. Since $F(M \cap H_\theta)$ is an open subset of $[\lambda]^\omega$ containing A_n there is $a_n \in [A_n]^{<\omega}$ such that $[a_n, A_n] \subseteq F(M \cap H_\theta)$. Let

$$ N_n = \mathrm{Sk}_{(H_{(2^\theta)^+}, \in, <_w)}(\{p_n, a_n, \mathcal{D}_n\}). $$

Then $N_n \cap H_\theta \in [a_n, A_n]$ and therefore $[a_n, N_n \cap H_\theta] \subseteq F(M \cap H_\theta)$. Working in N_n^*, we first find a countable elementary submodel X_n of H_θ containing a_n and then use the density of $\mathcal{D}_n$ to find $p_{n+1} \leq p_n \frown X_n$ in $\mathcal{D}_n \cap N_n$. Then for every model N in the difference $p_{n+1} \setminus p_n$, we have that $N \cap \lambda \in [a_n, N_n \cap H_\theta]$, and therefore, $N \cap \lambda \in F(M \cap H_\theta)$. This finishes the recursive step. It is clear that the sequence p_n $(n < \omega)$ can be chosen with the additional property that the models appearing in p_n's cover $M \cap H_\theta$, so if we define

$$ p = \left(\bigcup_{n < \omega} p_n \right) \cup \{ (\alpha, M \cap H_\theta) \}, $$

where $\alpha = \sup_{n<\omega} \alpha_{p_n}$, then the resulting sequence is a condition of our poset $\mathcal{P}$. Choose $q \leq p$ such that $C(q) \setminus \delta \neq \emptyset$. Then arguing as before we conclude that δ must belong to $C(q)$.   □

For $\alpha < \omega_1$, let

$$ \mathcal{D}_\alpha = \{ p \in \mathcal{P} : \alpha_p \geq \alpha \}. $$

It is clear that these sets are dense-open in $\mathcal{P}$ and that a filter that intersects them all will give us a continuous $\in$-chain of length ω_1 satisfying the conclusion of the Open Stationary Set-Mapping Reflection for the set-mapping F we started with.

This completes the proof.   □

Appendix A

Basic Notions

In this Appendix, we set out some of the standard notions and definitions that were used in these lecture notes.

A.1   Set theoretic notions

Definition. An *ordinal* is a transitive set well-ordered by $\in$. We denote the class of all ordinals by **ON**. Given ordinals α, β we say $\alpha < \beta$ iff $\alpha \in \beta$. Note that $\alpha = \{\beta \in \mathbf{ON} : \beta < \alpha\}$; i.e., each ordinal is equal (as a set) to the set of all of its predecessors.

Notation. Given a set X and a cardinal κ we denote:

$$[X]^\kappa = \{Y \subseteq X : |Y| = \kappa\};$$
$$[X]^{<\kappa} = \bigcup_{\lambda < \kappa} [X]^\lambda = \{Y \subseteq X : |Y| < \kappa\};$$
$$[X]^{\leq \kappa} = \bigcup_{\lambda \leq \kappa} [X]^\lambda = \{Y \subseteq X : |Y| \leq \kappa\}.$$

Notation. We will denote the set $[\mathbb{N}]^{<\omega}$ of all finite sets of natural numbers by *Fin*.

Definition. Given a p.o. $\langle \mathcal{P}, \leq \rangle$, by a *filter in* $\langle \mathcal{P}, \leq \rangle$ we mean a subset $\mathcal{F} \subseteq \mathcal{P}$ satisfying:

(i) $\mathcal{F} \neq \emptyset$, and

(ii) $p \leq q$ and $p \in \mathcal{F}$ implies $q \in \mathcal{F}$, and

(iii) for all $p, q \in \mathcal{F}$ there is an $r \in \mathcal{F}$ such that $r \leq p, q$.

Definition. Given a set X, by a *filter on X* we mean a family of subsets $\mathcal{F}$ of X satisfying:

(i) $\mathcal{F} \neq \emptyset$, and

(ii) $B \supseteq A \in \mathcal{F}$ implies $B \in \mathcal{F}$, and

(iii) $A \cap B \in \mathcal{F}$ for all $A, B \in \mathcal{F}$.

$\mathcal{F}$ is called an *ultrafilter on X* if, in addition, $\mathcal{F}$ satisfies:

(iv) $A \in \mathcal{F} \Leftrightarrow X \setminus A \notin \mathcal{F}$ for all $A \subseteq X$.

Fact. *A filter on a set X is precisely a filter in the p.o. $\langle \mathcal{P}(X) \setminus \{\emptyset\}, \subseteq \rangle$.*

Definition. Let $\langle \mathcal{P}, \leq \rangle$ be a p.o., and let $p, q \in \mathcal{P}$. We say that p, q are *compatible*, denoted $p \not\perp q$, if there is an $r \in \mathcal{P}$ such that $r \leq p, q$. Otherwise we say that p, q are *incompatible*, and write $p \perp q$.

Definition. Let $\langle \mathcal{P}, \leq \rangle$ be a p.o., and let $X \subseteq \mathcal{P}$.

(a) We call X *dense* in $\langle \mathcal{P}, \leq \rangle$ if $(\forall p \in \mathcal{P})(\exists q \in X)(q \leq p)$.

(b) We call X *open* in $\langle \mathcal{P}, \leq \rangle$ if $(\forall q \in X)(\forall p \in \mathcal{P})\ (p \leq q \rightarrow p \in X)$.

(c) We call X an *antichain* in $\langle \mathcal{P}, \leq \rangle$ if every two distinct elements of X are incompatible.

Definition. A p.o. $\langle \mathcal{P}, \leq \rangle$ is said to satisfy the *countable chain condition (c.c.c.)* if every antichain in $\langle \mathcal{P}, \leq \rangle$ is countable.

Definition. A p.o. $\langle \mathcal{P}, \leq \rangle$ is said to be *separative* if for each $p, q \in \mathcal{P}$ with $p \not\leq q$ there is an $r \leq p$ such that $r \perp q$.

Fact. *Let $\langle \mathcal{P}, \leq \rangle$ be an arbitrary p.o. Define the equivalence relation $\equiv$ of $\mathcal{P}$ by*

$$p \equiv q \qquad \Leftrightarrow \qquad (\forall r \in \mathcal{P})(r \perp q \leftrightarrow r \perp p).$$

Then $\leq$ naturally induces a partial order $\leq_\equiv$ on the quotient $\mathcal{P}/\equiv$, and the partial order $\langle \mathcal{P}/\equiv, \leq_\equiv \rangle$ is a separative p.o.

A.2 Δ-systems and free sets

The Δ-system Lemma and the Free-set Lemma have both shown to be useful in proofs that partial ordering satisfy the countable chain condition. So let us state them letting the reader supply their proofs or find them in the literature.

Theorem 110. *(The Δ-System Lemma) Every uncountable family $\mathcal{F}$ of finite sets contains an uncountable subfamily $\mathcal{F}_0$ forming a Δ-system, i.e., there is a finite set R such that for any distinct $F, G \in \mathcal{F}_0$,*

$$R = F \cap G.$$

The set R is called the root *of the Δ-system $\mathcal{F}_0$.*

Theorem 111. *(The Free-Set Lemma) Let $\{F_i\}(i \in I)$ be a family of sets such that there is a cardinal $\lambda < |I|$ such that $|F_i| < \lambda$ for all $i \in I$. Then there exists a subfamily $\{F_i\}(i \in J)$ for some $J \subseteq I$ such that $|J| = |I|$ and the subfamily is free, i.e., $i \notin F_j$ for all distinct $i, j \in J$.*

A.3 Topological notions

Definition. A subset U of a topological space X is said to be *regular-open* if $\text{Int}\,(\overline{U}) = U$.

Fact. *For any subset A of a topological space X, the set $\text{Int}\,(\overline{A})$ is regular-open in X.*

Definition. A topological space X is said to satisfy the *countable chain condition (c.c.c.)* if every family of pairwise disjoint open subsets of X is countable.

Notation. Let X be a topological space. We denote the family of all dense open subsets of X by $\mathcal{DO}(X)$.

The nature of Čech-complete spaces will be important in these lectures, and so we provide the following facts.

Definition. A $T_{3\frac{1}{2}}$-space X is called *Čech-complete* if it is a G_δ-subset of its Stone-Čech compactification βX.

Fact. *For any Čech-complete space X, we have $\mathfrak{m}(X) = \mathfrak{m}(\beta X)$, where βX denotes the Stone-Čech compactification of X.*

Proof. Suppose that $\{D_\alpha\}_{\alpha < \kappa}$ is a family of dense-open subsets of X with empty intersection. Note that $\omega < \kappa$. There are dense-open $\{U_n\}_{n < \omega}$ subsets of βX such that $\bigcap_{n < \omega} U_n = X$. For each $\alpha < \kappa$ there is a dense-open $E_\alpha \subseteq \beta X$ such that $E_\alpha \cap X = D_\alpha$. It easily follows that $\bigcap_{\alpha < \kappa} E_\alpha \cap \bigcap_{n < \omega} U_n = \emptyset$, and therefore $\mathfrak{m}(\beta X) \leq \mathfrak{m}(X)$.

Suppose that $\{E_\alpha\}_{\alpha < \kappa}$ is a family of dense-open subsets of βX with empty intersection. It easily follows that $\{E_\alpha \cap X\}_{\alpha < \kappa}$ is a family of dense-open subsets of X with empty intersection. Thus $\mathfrak{m}(X) \leq \mathfrak{m}(\beta X)$. $\quad\square$

Fact. *In any regular space X, the family of nonempty regular-open subsets forms a base for the topology on X.*

A.4   Boolean algebras

Definition. A *Boolean algebra* is a structure $\langle B, 0, 1, -, \vee, \wedge \rangle$ where B is a nonempty set, $0, 1 \in B$, $-$ is a unary operator on B, $\vee$ and $\wedge$ are binary operators on B, and which satisfies the following axioms:

(a) $x \wedge (y \wedge z) = (x \wedge y) \wedge z$.

(b) $x \wedge y = y \wedge x$.

(c) $x \vee (x \wedge y) = x$.

(d) $x \wedge (y \vee z) = (x \wedge y) \vee (x \wedge z)$.

(e) $x \wedge (-x) = 0$.

(a') $x \vee (y \vee z) = (x \vee y) \vee z$.

(b') $x \vee y = y \vee x$.

(c') $x \wedge (x \vee y) = x$.

(d') $x \vee (y \wedge z) = (x \vee y) \wedge (x \vee z)$.

(e') $x \vee (-x) = 1$.

Definition. Given a Boolean algebra B the relation $\leq$ on B defined by

$$x \leq y \quad \Leftrightarrow \quad x \vee y = y$$

is called the *Boolean order* on B.

Fact. *For any Boolean algebra B, the Boolean order on B is a partial order on B.*

Definition. Let B be a Boolean algebra. By a *filter* in B we mean a filter in $\langle B \setminus \{0\}, \leq \rangle$, where $\leq$ is the Boolean order on B. By an *ultrafilter* in B we mean a filter $\mathcal{U}$ in B additionally satisfying:

$$(\forall x \in B)\ (-x \in \mathcal{U} \leftrightarrow x \notin \mathcal{U}).$$

Fact. *Let B be a Boolean algebra. For any $x, y \in B$ we have*

$$x \vee y = \sup\{x, y\};$$
$$x \wedge y = \inf\{x, y\},$$

where the supremum and infimum are taken with respect to the Boolean order on B.

Definition. Let B be a Boolean algebra, and let $X \subseteq B$. We define

$$\bigvee X = \sup(X); \qquad \bigwedge X = \inf(X),$$

where the supremum and infimum are taken with respect to the Boolean order on B, provided that these exist. Otherwise we leave $\bigvee X$ and $\bigwedge X$ undefined.

Definition. A Boolean algebra B is called *complete* if $\bigvee X$ exists for all $X \subseteq B$.

Appendix B

Preserving Stationary Sets

B.1 Stationary sets

Definition. Let $A \subseteq \omega_1$.

1. A is called *closed* if for each increasing sequence $\{\alpha_n\}_{n\in\omega}$ of elements of A, $\sup_{n\in\omega} \alpha_n \in A$.

2. A is called *unbounded* if for each $\alpha < \omega_1$ there is a $\beta \in A$ with $\beta \geq \alpha$.

3. We call A a *club in* ω_1 if it is both closed and unbounded.

Definition. A subset $S \subseteq \omega_1$ is called *stationary* if it meets every club in ω_1.

Basic properties of these notions are best seen in a bit bigger generality.

Definition. Recall that $[\theta]^\omega$ is the collection of all countably infinite subsets of A. A subset S of $[\theta]^\omega$ is *unbounded* or *cofinal* if for every $a \in [\theta]^\omega$ there is $B \in S$ such that $A \subseteq B$. A subset C of $[\theta]^\omega$ is *closed* if $\bigcup \mathcal{X} \in C$ for every countable chain of elements of C. A subset S of $[\theta]^\omega$ is *stationary* if it intersects all closed and unbounded subsets of $[\theta]^\omega$.

Thus if $\theta = \omega_1$, the set $\{\alpha : \alpha < \omega_1\}$ is a closed and unbounded in $[\theta]^\omega$ and for $E \subseteq \omega_1$, we have that E is stationary in ω_1 if and only if $\{\alpha : \alpha \in E\}$ is stationary in $[\theta]^\omega$. Another way to generate closed and unbounded subsets of $[\theta]^\omega$ if to take an arbitrary map $f : \theta^{<\omega} \to \theta$ and consider

$$C_f = \{A \in [\theta]^\omega : f[A^{<\omega}] \subseteq A\}.$$

We have the following crucial properties of these notions.

Lemma. *For every closed and unbounded subset C of $[\theta]^\omega$ there is $f : \theta^{<\omega} \to \theta$ such that $C_f \subseteq C$.*

Lemma (Diagonal Intersection). *For every sequence C_ξ $(\xi < \theta)$ of closed and unbounded subsets of $[\theta]$, the set*

$$\triangle_{\xi<\theta} C_\xi = \{A \in [\theta]^\omega : (\forall \xi \in A) \ A \in C_\xi\}$$

is closed and unbounded in $[\theta]^\omega$.

Corollary (Pressing-Down Lemma). *Suppose S is a stationary subset of $[\theta]^\omega$ and $f : S \to \theta$ is a regressive map, i.e., $f(A) \in A$ for all $A \in S$. then f is constant on a stationary subset of S.*

B.2 Partial orders, Boolean algebras and topological spaces

In this section we will explain some correspondences between the three mathematical objects listed in the title.

Proposition. *Let $\langle \mathcal{P}, \leq \rangle$ be a partially ordered set. For each $x \in \mathcal{P}$, define*

$$\langle x \rangle = \{y \in \mathcal{P} : y \leq x\}.$$

Then the family

$$\mathcal{B} = \{\langle x \rangle : x \in \mathcal{P}\},$$

is a basis for a topology on $\mathcal{P}$. Denote the induced topology by $\mathcal{T}_\leq$.

Proof. Clearly $x \in \langle x \rangle$ for each $x \in \mathcal{P}$, and thus $\bigcup \mathcal{B} = \mathcal{P}$. Suppose that $y \in \langle x_1 \rangle \cap \langle x_2 \rangle$ for some $x_1, x_2 \in \mathcal{P}$. By transitivity it is easy to show that $y \in \langle y \rangle \subseteq \langle x_1 \rangle \cap \langle x_2 \rangle$. $\square$

Fact. *Let $\langle \mathcal{P}, \leq \rangle$ be a partial order. Given any $p \in \mathcal{P}$ we have*

$$Int\left(\overline{\langle p \rangle}\right) = \{q \in \mathcal{P} : (\forall r \leq q) \, r \not\perp p\}.$$

Our interest in separative partial orders extends from the following Fact:

Fact. *Let $\langle \mathcal{P}, \leq \rangle$ be a partial order. Then $\langle \mathcal{P}, \leq \rangle$ is separative iff $\langle p \rangle$ is a regular open set in $\langle \mathcal{P}, \mathcal{T}_\leq \rangle$ for each $p \in \mathcal{P}$.*

Proof. If $\langle \mathcal{P}, \leq \rangle$ is separative, then for each $p, q \in \mathcal{P}$ we have that $q \leq p$ iff $(\forall r \leq q)(r \not\perp p)$. It follows by the above Fact it follows that $q \in Int\left(\overline{\langle p \rangle}\right)$ iff $q \leq p$. Therefore $\langle p \rangle$ is regular-open in $\langle \mathcal{P}, \mathcal{T}_\leq \rangle$.

If $\langle \mathcal{P}, \leq \rangle$ is not separative, then there are $p, q \in \mathcal{P}$ such that $q \not\leq p$ and $(\forall r \leq q)(r \not\perp p)$. By the Fact above it follows that $q \in Int\left(\overline{\langle p \rangle}\right)$ but $q \notin \langle p \rangle$, and thus $\langle p \rangle$ is not regular-open in $\langle \mathcal{P}, \leq \rangle$. $\square$

Proposition. *Let X be a topological space. Denote by $RO(X)$ the family of all regular-open subsets of X. Define $-$, $\vee$, and $\wedge$ on $RO(X)$ as follows:*

$$-U = Int(X \setminus U),$$
$$U \vee V = Int(\overline{U \cup V}),$$
$$U \wedge V = U \cap V.$$

Then the structure $\langle RO(X), \emptyset, X, -, \vee, \wedge \rangle$ is a complete Boolean algebra, called the regular-open algebra *of X.*

Proof. Much of the proof relies either on simple set manipulations, or the fact that given any $A \subseteq X$, $Int(\overline{A})$ is a regular open subset of X.

To show that $RO(X)$ is complete, note that given $\{U_\alpha\}_{\alpha < \kappa} \subseteq RO(X)$, we have that

$$\bigvee_{\alpha < \kappa} U_\alpha = Int\left(\overline{\bigcup_{\alpha < \kappa} U_\alpha} \right). \qquad \square$$

Proposition. *Let B be a Boolean algebra. Denote by $Ult(B)$ the family of all ultrafilters in B. For each $x \in B$ define*

$$[x] = \{ \mathcal{U} \in Ult(B) : x \in \mathcal{U} \}.$$

Then the family

$$\mathcal{B} = \{ [x] : x \in B \}$$

is a basis for a topology on $Ult(B)$. Under the induced topology, $Ult(B)$ is called the Stone space *of B, denoted by $St(B)$.*

Proof. We trivially have that $\bigcup \mathcal{B} = B$. Given $\mathcal{U} \in [x] \cap [y]$, it is easy to show that $\mathcal{U} \in [x \wedge y] \subseteq [x] \cap [y]$. $\qquad \square$

Proposition. *Let $\langle \mathcal{P}, \leq \rangle$ be a partially ordered set, and let $K = St(RO(\langle \mathcal{P}, \mathcal{T}_\leq \rangle))$.*

(a) Given any dense-open set D in $\langle \mathcal{P}, \leq \rangle$, the set

$$D^* = \{ \mathcal{U} \in K : (\exists p \in D)(Int(\overline{\langle p \rangle}) \in \mathcal{U}) \}$$

is dense-open in K.

(b) Given any dense-open set E in K, the set

$$E_* = \{ p \in \mathcal{P} : [Int(\overline{\langle p \rangle})] \subseteq E \}$$

is dense-open in $\langle \mathcal{P}, \leq \rangle$.

Proof. (a) Clearly D^* is open in K, as for any $\mathcal{U} \in D^*$, if $p \in D$ it such that $\mathrm{Int}\,(\overline{\langle p \rangle}) \in \mathcal{U}$, then as $\mathrm{Int}\,(\overline{\langle p \rangle})$ is regular-open in $\langle \mathcal{P}, \mathcal{T}_\le \rangle$, it follows easily that $\mathcal{U} \in [\mathrm{Int}\,(\overline{\langle p \rangle})] \subseteq D^*$.

Given any nonempty $U \in \mathrm{RO}\,(\langle \mathcal{P}, \mathcal{T}_\le \rangle)$, as U is open in $\langle \mathcal{P}, \le \rangle$, it follows that $U \cap D \ne \emptyset$, so pick some $p \in U \cap D$. As $p \in U$, and U is regular open, it follows that $\mathrm{Int}\,(\overline{\langle p \rangle}) \subseteq U$, and therefore $[\mathrm{Int}\,(\overline{\langle p \rangle})] \subseteq [U]$. As $p \in D$, by definition it follows that $[\mathrm{Int}\,(\overline{\langle p \rangle})] \subseteq D^*$. Thus $D^* \cap [U] \ne \emptyset$, and thus D^* is dense in K.

(b) Suppose that $q \in E_*$, and $p \in \mathcal{P}$ with $p \le q$. Then we trivially have that $\langle p \rangle \subseteq \langle q \rangle$, and therefore $[\mathrm{Int}\,(\overline{\langle p \rangle})] \subseteq [\mathrm{Int}\,(\overline{\langle q \rangle})] \subseteq E$. Thus E_* is open in $\langle \mathcal{P}, \le \rangle$.

Given any $q \in \mathcal{P}$, as E is dense in K, it follows that $[\mathrm{Int}\,(\overline{\langle q \rangle})] \cap E \ne \emptyset$. As this set is also open, there is a $U \in \mathrm{RO}\,(\langle \mathcal{P}, \mathcal{T}_\le \rangle)$ such that $[U] \subseteq [\mathrm{Int}\,(\overline{\langle q \rangle})] \cap E$. Pick an arbitrary $r \in U$. Then we have

$$[\mathrm{Int}\,(\overline{\langle r \rangle})] \subseteq [U] \subseteq [\mathrm{Int}\,(\overline{\langle q \rangle})] \cap E.$$

It follows immediately that $r \in E_*$. We also have $r \in \mathrm{Int}\,(\overline{\langle q \rangle})$, and therefore $r \not\perp q$, so there is a $p \in \mathcal{P}$ with $p \le q, r$. As $r \in E_*$, and by the above E_* is open, it follows that $p \in E_*$. Thus E_* is dense in $\langle \mathcal{P}, \le \rangle$. $\qquad \square$

We will now provide for a category-type number for partial orders that will be closely related to the category number of the topological spaces. Unfortunately, the one defined below will not be the most useful in forcing-type arguments, though for particularly nice partial orders, a forcing-capable definition will also work. This will be better explained in the sequel.

Definition. Let $\langle \mathcal{P}, \le \rangle$ be a partial order. We define its *quasi-category number* as follows:

$$\tilde{\mathrm{m}}(\langle \mathcal{P}, \le \rangle) = \min \left\{ |\mathcal{D}| : \begin{array}{l} \mathcal{D} \text{ is a family of dense-open subsets of } \mathcal{P} \text{ and there} \\ \text{is no centred subset of } \mathcal{P} \text{ which meets each set in } \mathcal{D} \end{array} \right\}$$

Proposition. *Let $\langle \mathcal{P}, \le \rangle$ be a partially ordered set, and let $K = St(RO(\langle \mathcal{P}, \mathcal{T}_\le \rangle))$. Then $\mathrm{m}(K) = \tilde{\mathrm{m}}(\langle \mathcal{P}, \le \rangle)$.*

Proof.

- Suppose that $\tilde{\mathrm{m}}(\langle \mathcal{P}, \le \rangle) \le \kappa$, and let $\{D_\alpha\}_{\alpha < \kappa}$ be a family of dense-open subsets of $\langle \mathcal{P}, \le \rangle$ such that no centred subset of $\langle \mathcal{P}, \le \rangle$ meets them all. Consider the family $\{D_\alpha^*\}_{\alpha < \kappa}$ of dense-open subsets of K. Suppose that $\mathcal{U} \in \bigcap_{\alpha < \kappa} D_\alpha^*$. Note that the family

$$\mathcal{A} = \{p \in \mathcal{P} : \mathrm{Int}\,(\overline{\langle p \rangle}) \in \mathcal{U}\}$$

is a centred set in $\langle \mathcal{P}, \leq \rangle$.

For any $\alpha < \kappa$, as $\mathcal{U} \in D_\alpha^*$, there is a $p \in D_\alpha$ such that $\mathrm{Int}\,(\overline{\langle p \rangle}) \in \mathcal{U}$. Then, by definition of $\mathcal{A}$ it follows that $p \in \mathcal{A}$, and thus $\mathcal{A} \cap D_\alpha \neq \emptyset$. Thus, $\mathcal{A}$ meets each of the D_α's, which is a contradiction. Therefore $\bigcap_{\alpha<\kappa} D_\alpha^* = \emptyset$, and therefore $\mathfrak{m}(K) \leq \kappa$.

- Suppose that $\mathfrak{m}(K) \leq \kappa$, and $\{E^{(\alpha)}\}_{\alpha<\kappa}$ is a family of dense-open sets in K with empty intersection. Consider the family $\{E_*^{(\alpha)}\}_{\alpha<\kappa}$ if dense-open sets in $\langle \mathcal{P}, \leq \rangle$. Suppose that $\mathcal{A}$ is a centred set in $\langle \mathcal{P}, \leq \rangle$ meeting each $E_*^{(\alpha)}$.

  It trivially follows that $\{\mathrm{Int}\,(\overline{\langle p \rangle}) : p \in \mathcal{A}\}$ is a centred set in $\mathrm{RO}\,(\langle \mathcal{P}, \mathcal{T}_\leq \rangle)$, and therefore there is an ultrafilter $\mathcal{U}$ containing this set.

  For each $\alpha < \kappa$, as $\mathcal{A} \cap E_*^{(\alpha)} \neq \emptyset$, there is some p in the intersection. As $p \in \mathcal{A}$, by definition we have that $\mathrm{Int}\,(\overline{\langle p \rangle}) \in \mathcal{U}$. As $p \in E_*^{(\alpha)}$, be definition it follows that $\mathcal{U} \in [\mathrm{Int}\,(\overline{\langle p \rangle})] \subseteq E^{(\alpha)}$, and therefore $\mathcal{U} \in E^{(\alpha)}$. It immediately follows that $\mathcal{U} \in \bigcap_{\alpha<\kappa} E^{(\alpha)} = \emptyset$, which is clearly a contradiction. Therefore there is no centred set in $\langle \mathcal{P}, \leq \rangle$ meeting each $E_*^{(\alpha)}$, and thus $\tilde{\mathfrak{m}}(\langle \mathcal{P}, \leq \rangle) \leq \kappa$. $\qquad\square$

Since the definition of $\tilde{\mathfrak{m}}(\langle \mathcal{P}, \leq \rangle$ mentions centred sets, as opposed to filters, it is not quite as useful for forcing arguments. However, a simple change by replacing "centred subset" to "filter" will not work for partial orders in general. There are, however, certain classes of partial orders where this change is useful, and therefore we will make the following definition:

Definition. Given a partially ordered set $\langle \mathcal{P}, \leq \rangle$, we define its *category number* as follows:

$$\mathfrak{m}(\langle \mathcal{P}, \leq \rangle) = \min\left\{ |\mathcal{D}| : \begin{array}{l} \mathcal{D} \text{ is a family of dense-open subsets of } \mathcal{P} \text{ and} \\ \text{there is no filter on } \mathcal{P} \text{ which meets each set in } \mathcal{D} \end{array} \right\}.$$

Notation.   1. $\mathfrak{m}_{\mathrm{po}} = \min\{\mathfrak{m}(\langle \mathcal{P}, \leq \rangle) : \langle \mathcal{P}, \leq \rangle \text{ has the c.c.c.}\}$.

2. $\mathfrak{m}_{\mathrm{po}\leq\kappa} = \min\{\mathfrak{m}(\langle \mathcal{P}, \leq \rangle) : \langle \mathcal{P}, \leq \rangle \text{ has the c.c.c., and } |\mathcal{P}| \leq \kappa\}$.

Proposition. *Let $\langle \mathcal{P}, \leq \rangle$ be a partially ordered set, and let $K = St(RO(\langle \mathcal{P}, \mathcal{T}_\leq \rangle))$.*

(i) $K \in \mathbf{CCC}$ iff $\langle \mathcal{P}, \leq \rangle$ has the c.c.c.

(ii) If $\langle \mathcal{P}, \leq \rangle$ has the c.c.c., then $\mathfrak{m}(\langle \mathcal{P}, \leq \rangle) = \mathfrak{m}(K)$.

Proof.   (i) Let $\{U_\alpha\}_{\alpha<\omega_1}$ is a family of pairwise disjoint open subsets of K. Without loss of generality, each U_α is a basic open set, i.e., $U_\alpha = [V_\alpha]$ for some $V_\alpha \in \mathrm{RO}\,(\langle \mathcal{P}, \mathcal{T}_\leq \rangle)$. For each $\alpha < \omega_1$ pick some $p_\alpha \in V_\alpha$.

Note that $\text{Int}\,(\overline{\langle p_\alpha \rangle}) \subseteq V_\alpha$. Suppose that $p_\alpha \not\perp p_\beta$. Then there is a $q \in \mathcal{P}$ with $q \leq p_\alpha, p_\beta$. This immediately implies that

$$q \in \langle q \rangle \subseteq \text{Int}\,(\overline{\langle q \rangle}) \subseteq V_\alpha \cap V_\beta.$$

It then follows that $[\text{Int}\,(\overline{\langle q \rangle})] \subseteq [V_\alpha] \cap [V_\beta]$, which is a contradiction. Thus, $\{p_\alpha\}_{\alpha < \omega_1}$ is an antichain in $\langle \mathcal{P}, \leq \rangle$.

Let $\{p_\alpha\}_{\alpha < \omega_1}$ be an antichain in $\langle \mathcal{P}, \leq \rangle$. For each $\alpha < \omega$ set $V_\alpha = \text{Int}\,(\overline{\langle p_\alpha \rangle})$ in the topological space $\langle \mathcal{P}, \mathcal{T}_\leq \rangle$. Consider now the family $\{[V_\alpha]\}_{\alpha < \omega_1}$ in K. If $\mathcal{U} \in [V_\alpha] \cap [V_\beta]$ for some $\alpha \neq \beta < \omega_1$, it follows that $V_\alpha \cap V_\beta \neq \emptyset$, and so pick some $q \in V_\alpha \cap V_\beta = \text{Int}\,(\overline{\langle p_\alpha \rangle}) \cap \text{Int}\,(\overline{\langle p_\beta \rangle})$. By an above fact it immediately follows that $p_\alpha \not\perp p_\beta$.

(ii) The proof of this is similar and therefore left to the interested reader.

$\square$

B.3　A topological translation of stationary set preserving

In this section we will translate the notion of preserving stationary sets into a topological property of Čech-complete spaces. We will first give an equivalent condition that a Čech-complete space has the c.c.c.

Fact. *A Čech-complete space K satisfies the c.c.c. iff for every family $\{O_\xi\}_{\xi < \omega_1}$ of regular-open subsets of K there is a club $C \subseteq \omega_1$ such that*

$$\mathcal{O}_{[\beta, \gamma)} = \text{Int}\left(\overline{\bigcup_{\beta \leq \xi < \gamma} O_\xi} \right)$$

is a constant sequence for all $\beta < \gamma$ with $\beta, \gamma \in C$.

Theorem 112. *Let K be a given Čech-complete space. Then the following are equivalent:*

1. *Every dense subposet $\mathcal{P}$ of $\mathcal{R} = \mathcal{R}(K) = RO(K) \setminus \{\emptyset\}$ is stationary set preserving.*

2. *For every sequence $\{O_\xi\}_{\xi < \omega_1}$ of nonempty regular open subsets of K and each nonempty open $U \subseteq K$ either*

   (a) *there is a nonempty open $V \subseteq U$ such that the set*

   $$\{\xi < \omega_1 : V \cap O_\xi \neq \emptyset\}$$

   is countable, or

(b) there is a club $\Gamma \subseteq \omega_1$ such that $Int\left(\overline{U \cap \bigcap_{\beta<\gamma} \mathcal{O}_{(\beta,\gamma]}}\right) \neq \emptyset$ for each $\gamma \in \Gamma$.

Proof.

(1 ⇒ 2) Suppose that every dense subposet of $\mathcal{R} = \mathrm{RO}\,(K) \setminus \{\emptyset\}$ is stationary set preserving. Let $\{O_\xi\}_{\xi<\omega_1}$ be a family of regular-open subsets of K, and U a given nonempty open subset of K. Without loss of generality we may assume that U is regular open.

Suppose that the set $\{\xi < \omega_1 : V \cap O_\xi \neq \emptyset\}$ is uncountable for each nonempty open $V \subseteq U$. For each $p \in \mathcal{R}$ we will define $C(p) \subseteq \omega_1$ as follows:

- If $p \subseteq U$ then
    - for each $\xi < \omega_1$, $\xi + 1 \in C(p)$ iff $p \subseteq \overline{O_\xi}$, and
    - for each limit $\gamma < \omega_1$, $\gamma \in C(p)$ iff $p \subseteq \overline{p \cap \bigcap_{\beta<\gamma} \mathcal{O}_{(\beta,\gamma]}}$.
- If $p \not\subseteq U$ and $p \cap U \neq \emptyset$, then $C(p) = C(U \cap p)$.
- If $p \cap U = \emptyset$, then $C(p) = \omega_1$.

Claim 112.1. $\langle C(p) : p \in \mathcal{R} \rangle$ *is a* $\mathcal{R}$-club.

Proof of claim. By definition it is clear that $\mathcal{R}$-monotonicity holds. Our assumption that $\{\xi < \omega_1 : V \cap O_\xi \neq \emptyset\}$ is uncountable for each nonempty open $V \subseteq U$ ensures that $\mathcal{R}$-unboundedness holds below U. As $C(p) = \omega_1$ for p disjoint from U, it follows easily that $\mathcal{R}$-unboundedness holds in general.

To show $\mathcal{R}$-closedness, we may take, without loss of generality, some $p \in \mathcal{R}$ with $p \subseteq U$. Suppose that $\xi + 1 \notin C(p)$. Then, by definition, $p \not\subseteq \overline{O_\xi}$, and therefore $p \setminus \overline{O_\xi}$ is a nonempty open set. Thus there is some $q \in \mathcal{R}$ with $q \subseteq p \setminus \overline{O_\xi}$. It trivially follows that for all $r \leq q$ we have $r \cap \overline{O_\xi} = \emptyset$, and therefore $C(r) \cap (\xi, \xi + 1] = \emptyset$.

Suppose that γ is a limit ordinal not in $C(p)$. Then there is a $q \in \mathcal{R}$ with $q \subseteq p \setminus \overline{p \cap \bigcap_{\beta<\gamma} \mathcal{O}_{(\beta,\gamma]}}$. Note that $q \cap \bigcap_{\beta<\gamma} \mathcal{O}_{(\beta,\gamma]} = \emptyset$. Suppose now that $q \cap \mathcal{O}_{(\beta,\gamma]}$ is dense in q for each $\beta < \gamma$. It then follows by the Baire Category Theorem that $q \cap \bigcap_{\beta<\gamma} \mathcal{O}_{(\beta,\gamma]}$ is also dense in q. However, this clearly contradicts the fact that q and $\bigcap_{\beta<\gamma} \mathcal{O}_{(\beta,\gamma]}$ are disjoint.

Thus, there is a $\beta < \gamma$ such that $q \cap \mathcal{O}_{(\beta,\gamma]}$ is not dense in q. Then there is an $r \leq q$ with $r \subseteq q \setminus \overline{q \cap \mathcal{O}_{(\beta,\gamma]}}$. It trivially follows that $r \cap O_\xi = \emptyset$ for each $\beta < \xi \leq \gamma$. Take any $s \leq r$. It is trivial to check that $\xi + 1 \notin C(s)$ for all $\beta < \xi < \gamma$. For any limit $\beta < \delta \leq \gamma$ we have that $s \cap \mathcal{O}_{(\mu,\delta]} = \emptyset$ for each $\mu < \delta$, and therefore it follows that $\delta \notin C(s)$. $\square$

Let $S = \{\gamma < \omega_1 : \mathrm{Int}\left(\overline{U \cap \bigcap_{\beta<\gamma} \mathcal{O}_{(\beta,\gamma]}}\right) = \emptyset\}$.

Claim 112.2. *S is not stationary.*

Proof of claim. If S is stationary, then there is a $q \subseteq U$ in $\mathcal{R}$ such that $C(q) \cap S \neq \emptyset$. Without loss of generality, let $\gamma \in C(q) \cap S$ be a limit ordinal. Then by definition it follows that

$$q \subseteq \mathrm{Int}\left(\overline{q \cap \bigcap_{\beta<\gamma} \mathcal{O}_{(\beta,\gamma]}}\right) \subseteq \mathrm{Int}\left(\overline{U \cap \bigcap_{\beta<\gamma} \mathcal{O}_{(\beta,\gamma]}}\right) = \emptyset$$

which is clearly a contradiction.   □

As S is not stationary, then there is a club $\Gamma \subseteq \omega_1$ disjoint from S. Without loss of generality we may assume that Γ is a set of limit ordinals. Then for each $\gamma \in \Gamma$, as $\gamma \notin S$ by definition we have

$$\mathrm{Int}\left(\overline{U \cap \bigcap_{\beta<\gamma} \mathcal{O}_{(\beta,\gamma]}}\right) \neq \emptyset.$$

Thus condition (2b) holds, as desired.

(2 $\Rightarrow$ 1) Let $\mathcal{P}$ be a dense subposet of $\mathrm{RO}(X) \setminus \{\emptyset\}$, let $\langle C(p) : p \in \mathcal{P}\rangle$ be a $\mathcal{P}$-club, and let $S \subseteq \omega_1$ be stationary. Let $\bar{p} \in \mathcal{P}$ be an arbitrary condition. For each $\xi < \omega_1$ define

$$O_\xi = \mathrm{Int}\left(\overline{\bigcup\{p \in \mathcal{P} : \xi \in C(p)\}}\right).$$

Then either

(a) There is a nonempty $V \subseteq \bar{p}$ such that the set $\{\xi < \omega_1 : V \cap O_\xi \neq \emptyset\}$ is countable, or

(b) there is a club $\Gamma \subseteq \omega_1$ such that $\mathrm{Int}\left(\overline{\bar{p} \cap \bigcap_{\beta<\gamma} \mathcal{O}_{(\beta,\gamma]}}\right) \neq \emptyset$ for all $\gamma \in \Gamma$.

Note that by $\mathcal{P}$-unboundedness, (a) cannot hold, and therefore let $\Gamma \subseteq \omega_1$ be a club witnessing (b). Choose some arbitrary $\gamma \in \Gamma \cap S$. Note that as $\mathrm{Int}\left(\overline{\bar{p} \cap \bigcap_{\beta<\gamma} \mathcal{O}_{(\beta,\gamma]}}\right)$ is a nonempty regular-open set, then there is a $q \in \mathcal{P}$ with $q \leq \bar{p}$ and $q \subseteq \mathrm{Int}\left(\overline{\bar{p} \cap \bigcap_{\beta<\gamma} \mathcal{O}_{(\beta,\gamma]}}\right)$.

Claim 112.3. $\gamma \in C(q)$.

Proof of claim. If $\gamma \notin C(q)$, then by $\mathcal{P}$-closedness there is a $\beta < \gamma$ and a $r \leq q$ such that $C(s) \cap (\beta, \gamma] = \emptyset$ for all $s \leq r$. Note that as $r \leq q$, by choice of q it follows that $r \subseteq \mathrm{Int}\left(\overline{\overline{p} \cap \bigcap_{\beta < \gamma} \mathcal{O}_{(\beta, \gamma]}}\right) \subseteq \mathcal{O}_{(\beta, \gamma]}$. It then follows that there is a $\beta < \xi \leq \gamma$ with $r \cap O_\xi \neq \emptyset$, and since this is a regular-open subset of K, there is an $s \leq r$ with $s \subseteq r \cap O_\xi$. By definition of O_ξ it follows that for some $p \in \mathcal{P}$ with $\xi \in C(p)$ we have $s \cap p \neq \emptyset$, and thus there is a $t \leq s$ with $t \subseteq s \cap p$. By $\mathcal{P}$-monotonicity it follows that $\xi \in C(t)$, and thus $C(t) \cap (\beta, \gamma] \neq \emptyset$, contradicting our choice of r. $\qquad \square$

It then follows that $\mathcal{P}$ is stationary set preserving. $\qquad \square$

Notation. Let $\mathbf{M}$ denote the class of all Čech-complete spaces satisfying either (hence both) conditions of the above Lemma.

Corollary 113. $\mathrm{mm} \leq \mathrm{m}(\mathbf{M})$.

Proof. Let $K \in \mathbf{M}$, and let $\{D_\alpha\}_{\alpha < \kappa}$ be a family of dense-open subsets of K with $\kappa < \mathrm{mm}$. Take $\mathcal{P} = \mathrm{RO}\,(K) \setminus \{\emptyset\}$. By the above Lemma, $\mathcal{P}$ is stationary set preserving.

For each $\alpha < \omega_1$ define

$$E_\alpha = \{p \in \mathcal{P} : p \subseteq D_\alpha\}.$$

Clearly $\{E_\alpha\}_{\alpha < \kappa}$ is a family of dense-open subsets of $\mathcal{P}$, and thus there is a filter $\mathcal{G} \subseteq \mathcal{P}$ intersecting each E_α. $\qquad \square$

Fact. $\mathrm{mm} = \mathrm{m}(\mathbf{M})$.

Proof. It remains to show the inequality $\mathrm{mm} \geq \mathrm{m}(\mathbf{M})$. We leave this as an exercise to the interested reader. $\qquad \square$

Appendix C

Historical and Other Comments

We recommend the reader the article [9] of Gilles Godefroy that surveys the work and life of René Baire as well as some of the early mathematical uses of the Baire Category Theorem.

The method of *forcing* was created in the 1960's by P. Cohen to solve the independence problem of the Continuum Hypothesis. This method involves extending a transitive model M of ZFC (called the *ground model*) by adding a new *generic* set G, and closing off under the usual set-theoretic operations to construct a larger model denoted $M[G]$. This is a metamathematical method, as it involves the construction of new models to prove independence results.

Internal forcing was byproduct of the work by R. Solovay and S. Tennenbaum [25] where a first instance of an iterated forcing was used in order to prove the consistency of the *Souslin hypothesis*. The internal forcing itself has been made explicit in the paper of D.A. Martin and R. Solovay [17] who were to isolate Martin's axiom, the first of the Strong Baire Category Principles. In our notation their MA_{ω_1} is written as $\mathfrak{m} > \omega_1$.

The Martin Maximum, MM, which in our notation is stated as $\mathfrak{mm} > \omega_1$, was introduced by M. Foreman, M. Magidor, and Shelah[7] as a natural extension of the Proper Forcing Axiom and the Semi-Proper Forcing Axiom introduced earlier by J. Baumgartner and S. Shelah (see [21]).

The method of using elementary submodels as side conditions for building proper poset was introduced by the author in the early 1980's (see, for example, [27]) as a byproduct of the solution of the S-space problem from topology. Since then it has been used in essentially all major advances of this area many of which we chose to reproduce above. In more recent years this side-condition method has been lifted by Mitchell, Friedman and Neeman to higher cardinals where a new generation of Forcing Axioms is slowly emerging. There is no doubt that these new Forcing Axiom will be as rich in consequences. This, however, calls for a theory of building posets with the higher side conditions that would reach applications not in the realm of Forcing Axioms considered in these sets of notes.

The first systematic study of the Strong Baire Category Principles from the point of view of their applications is Fremlin's book [8] which we strongly recommend to the reader. In fact these lecture notes is an attempt to continue this line of presentation of the strong Baire category principles. However at this stage in the construction of these notes, we make no attempt for a proper in-depth historical analysis of results that we present. We include some references where the reader can find the origins for some of the result we present here but a more complete reference list is clearly lacking.

Bibliography

[1] B. Balcar, T. Jech, *Weak distributivity, a problem of von Neumann and the mystery of measurability.* Bull. Symbolic Logic 12 (2006), no. 2, 241–266.

[2] J. Baumgartner, *All $\aleph_1$-dense sets of reals can be isomorphic.* Fund. Math. 79 (1973), no. 2, 101–106.

[3] M. Bekkali, *Topics in set theory. Lebesgue measurability, large cardinals, forcing axioms, rho-functions.* Notes on lectures by Stevo Todorcevic. Lecture Notes in Mathematics, 1476. Springer-Verlag, Berlin, 1991.

[4] H. G. Dales and W. H. Woodin, *An introduction to independence for analysts.* London Mathematical Society Lecture Note Series, 115. Cambridge University Press, Cambridge, 1987. xiv+241 pp.

[5] R. Engelking, *General topology.* Second edition. Sigma Series in Pure Mathematics, 6. Heldermann Verlag, Berlin, 1989. viii+529 pp.

[6] P. Erdős, A. Hajnal, A. Máté and R. Rado, *Combinatorial set theory: Partition Calculus for Cardinals.* North Holland, Amsterdam 1984.

[7] M. Foreman, M. Magidor, S. Shelah, *Martin's maximum, saturated ideals and nonregular ultrafilters*, Ann. Math. **127** (1988), 1–47.

[8] D. H. Fremlin, *Consequences of Martin's Axiom*, Camb. Univ. Press, Cambridge, 1984.

[9] G. Godefroy, *Le Lemme de Baire,* www.dma.ens.fr/culturemath/maths/pdf/analyse/baire.pdf

[10] P. Hajek, V. Montesinos, J. Vanderwerff, and V. Zizler, *Biorthogonal systems in Banach spaces*, Canadian Mathematical Society Books in Mathematics, Springer 2008.

[11] T. Jech, *Set theory*. The third millennium edition, revised and expanded. Springer Monographs in Mathematics. Springer-Verlag, Berlin, 2003. xiv+769 pp.

[12] A. Kechris, *Classical Descriptive Set Theory*, Graduate Texts in Mathematics, 156, Springer Verlag, 1995.

[13] P. Koszmider, *Models as side conditions*, Techniques and Applications of Set Theory, Centre de Recerca Matemàtica, vol. 8, December 1996, pp. 19–28.

[14] K. Kunen, *Set Theory*, North Holland 1980.

[15] K. Kuratowski, *Topologie, vol.I*, Academic Press, 1996.

[16] J. Lindenstrauss, L. Tzafriri, *Classical Banach spaces, I*, Springer-Verlag, Berlin 1977.

[17] D. A. Martin and R. M. Solovay, Internal Cohen extensions, *Ann. Math. Logic* **2** (1970), pp. 143–178.

[18] J. T. Moore, *The proper forcing axiom.* Proceedings of the International Congress of Mathematicians. Volume II, 329, Hindustan Book Agency, New Delhi, 2010.

[19] P. Nyikos, L. Soukup, and B. Velickovic, *Hereditary normality of γN-spaces.* Topology Appl. 65 (1995), no. 1, 9–19.

[20] F. Rothberger, On families of real functions with a denumerable base, *Ann. Math.* **45** (1944), pp. 397–406.

[21] S. Shelah *Proper and improper forcing*, Springer 1998.

[22] W. Sierpinski, *Lhypothèse généralisée du continu et l'axiome du choix*, Fundamenta Mathematicae, **34**, no. 1, 1947, p. 1–5.

[23] W. Sierpiński, *Cardinal and Ordinal Numbers*, Państwowe Wydawnictwo Naukowe, Warszawa, 1958.

[24] I. Singer, *Bases in Banach spaces II*, Springer-Verlag, Berlin 1981.

[25] R. M. Solovay and S. Tennenbaum, Iterated Cohen extensions and Souslin's problem, *Ann. Math.* **94** (1971), pp. 201–245.

[26] S. Todorcevic, *Trees and linearly ordered sets*, In: Handbook of set-theoretic topology, 235–293, North-Holland, Amsterdam, 1984.

[27] S. Todorcevic *A note of the Proper Forcing Axiom*, In: Contemporary Math., Vol 31 (1984), 209–218. American Mathematical Society, Providence.

[28] S. Todorcevic, *Partition Problems in Topology*, Contemporary Mathematics, vol. 84, American Mathematical Society, 1989, Providence, Rhode Island.

[29] S. Todorcevic *A classification of transitive relations on ω_1. Proc. London. Math. Soc.*(3) **73** (1996), pp. 501–533.

[30] S. Todorcevic, *Chain-condition methods in topology*, Topology and its Appl. **101** (2000), 45–82.

[31] S. Todorcevic, *Biorthogonal systems and quotient spaces via Baire Category methods* Math. Annalen 335 (2006), 687–715.

[32] S. Todorcevic, *Walks on ordinals and their characteristics*. Birkauser, Basel 2007.

[33] S. Todorcevic, *Combinatorial dichotomies in set theory*. Bull. Symbolic Logic 17 (2011), no. 1, 1–72.

[34] J. W. Tukey, Convergence and Uniformity in Topology, Annals of Mathematics Studies, No. 2, Princeton University Press, 1940.

[35] N. H. Williams, Combinatorial set theory, North Holland, Amsterdam 1977.